Study Guide

for

Kendall's

Sociology In Our Times
The Essentials

Sixth Edition

Kathryn Sinast Mueller
Baylor University

Australia • Brazil • Canada • Mexico • Singapore • Spain • United Kingdom • United States

Printed in the United States of America

1 2 3 4 5 6 7 10 09 08 07

Printer: Thomson West

Cover Image: © Todd Berman, *The Mission Apartments*

ISBN-13: 978-0-495-09942-0
ISBN-10: 0-495-09942-2

Thomson Higher Education
10 Davis Drive
Belmont, CA 94002-3098
USA

For more information about our products, contact us at:
Thomson Learning Academic Resource Center
1-800-423-0563

For permission to use material from this text or product, submit a request online at **http://www.thomsonrights.com.**
Any additional questions about permissions can be submitted by email to **thomsonrights@thomson.com.**

DEDICATED TO

Kenneth R. Mueller

The love of my life and the
"wind beneath my wings"

WITH HEARTFELT THANKS TO

Kristina Cramb
Jennifer Amatya

Whose dedication, skill, and dependability
to this publication made all this possible

TABLE OF CONTENTS

PREFACE

Welcome to the Study Guide for **Sociology in Our Times: The Essentials**, 6th Edition. The text, in its sixth edition, is even more outstanding than the first edition! This Study Guide has been written to assist you in mastering the material in your text and enhancing your grade on your instructor's tests. Do not use it as a replacement for the text; it does not contain all of the passages from the text, nor does it depict all of the major topics illustrating the sociological concepts --your text does that! Each chapter in the Study Guide corresponds with the same chapter in **Sociology in Our Times: The Essentials**, 6th Edition, and has the following components:

Brief Chapter Outline

Chapter Summary

Learning Objectives

Key Terms

Key People

Chapter Outline

Analyzing and Understanding the Boxes

Practice Tests

Sociology in Our Times: Diversity Issues

Internet Resources and Exercises

InfoTrac Online Readings and Exercises

Student Class Projects and Activities

How To Use This Study Guide

1. Before reading each chapter, look over the Brief Chapter Outline and Chapter Summary to get an overview of the topics in each chapter.

2. As you read, use the Learning Objectives, Key Terms, and Key People as a guide for taking notes. (You may wish to start a sociology notebook with concise responses for study purposes.)

3. Look for recurring themes in each chapter and between chapters. For example, how do functionalist, conflict, and symbolic interactionist perspectives view each of the topics being examined? Outline the key components of each theory in your notebook.

4. Look at the topics and issues presented through the lenses of race, class, gender, age, and (where applicable) ability/disability.

5. In the section on "Analyzing and Understanding the Boxes," outline the key points and possible discussion questions for each of the primary boxes -- "Sociology and Everyday Life," and "You Can Make a Difference" -- that are included in each chapter of the textbook. Next, examine any secondary boxes in each chapter of the textbook and respond to each specific question.

6. After you have read the chapter, take the practice test under actual testing situations (i.e., "closed book," sitting at a table, and timing yourself). Grade your practice test using the answers at the end of each Study Guide chapter. Use the page numbers indicated on the practice test and in the answers to revisit the text for your information on each item you missed. (You may wish to save the practice tests to use as a self test before you take an in-class examination.)

7. Using the Internet Resources and Exercises will provide you with useful information when doing research, or when you want to focus on a specific term or idea featured in the text. Experience and enjoy the exercises! You will enjoy learning more about Sociology!

8. Access the InfoTrac College Edition online readings. Short answer and discussion questions are available at each site. These readings are easy to search and find and are fun to read. Each InfoTrac article corresponds with important concepts from each particular chapter. The articles selected range from those found in major sociological journals, to other leading periodicals.

9. The student class projects and activities are designed to involve you actively in the learning process and to help you recognize the relevance of sociology to your life and to our global society. Some of these projects may be assigned as a part of the course requirements.

Best wishes as you explore the text *Sociology in Our Times*: *The Essentials*, 6th Edition, and as you use this study guide. May you experience and enjoy the sociological imagination! We invite your questions, comments or suggestions. You may contact us at this address:

Kathryn Sinast Mueller
Baylor University
P.O. Box 97326
Waco, Texas 76798-7326
e-mail address:
Kathryn_Mueller@baylor.edu

Diana Kendall
Baylor University
P.O. Box 97326
Waco, Texas 76798-7326
email address:
Diana_Kendall@baylor.edu

From Students to Students about the Study Guide

Dear students,
This study guide is a comprehensive and detailed supplemental source to Kendall's *Sociology in Our Times: The Essentials, Sixth Edition.* Professor Mueller has done a superb job in outlining both the broad and detailed sociological concepts that are found in this textbook. Each individual chapter in this particular study guide carefully examines the overall content of the main text. There is, in fact, a direct correlation between the study guide chapters and the textbook chapters. I recommend this guide as a supplement for any student's Introduction to Sociology course. This guide is a wonderful tool when one is reviewing class lectures, presentations, semester projects, "understanding the boxes," or any other information that might be pertinent to this class. It is especially helpful during "crunch time," the time right before a big examination. This source has helped me and I know that it will do the same for you!

Kristina Cramb
Baylor University, Class of 2005
B.A. Politcal Science, M.A. 2008
Gender Studies Minor

Dear students,
I recommend reading and studying from this supplemental study guide. This source is complementary to the text. The internet activities, which are found at the end of each chapter, are both fun and instructive. As a side note, I just wanted to let the students know that I have helped Professor Mueller grade the student submitted projects, and they were awesome! This is a wonderful way to learn the required material.

Kacy Sandidge
Sociology Major
Baylor University, Class of 2007

Dear students,
Professor Mueller has put so much time and effort into creating this study guide to help students understand the sociological principles found in Kendall's *Sociology in Our Times: The Essentials*. I could tell you about all the long nights we spent in the office compiling this supplemental text. Put it to good use and make our efforts worthwhile. I think you will find it to be a very helpful study tool. Good luck!

Jennifer R. Amatya
Baylor University, Class of 2007
B.A. International Studies
Church-State Studies minor

CHAPTER 1
THE SOCIOLOGICAL PERSPECTIVE AND RESEARCH PROCESS

BRIEF CHAPTER OUTLINE

Putting Social Life into Perspective
- Why Study Sociology?
- The Sociological Imagination
- The Importance of a Global Sociological Imagination

The Development of Sociological Thinking
- Early Thinkers: A Concern with Social Order and Stability
- Differing Views on the Status Quo: Stability Versus Change
- The Beginnings of Sociology in the United States

Contemporary Theoretical Perspectives
- Functionalist Perspectives
- Conflict Perspectives
- Symbolic Interactionist Perspectives
- Postmodern Perspectives

The Sociological Research Process
- The "Conventional" Research Model
- A Qualitative Approach to Researching Suicide

Research Methods
- Survey Research
- Secondary Analysis of Existing Data
- Field Research
- Experiments

Ethical Issues in Sociological Research

CHAPTER SUMMARY

Sociology is the systematic study of human society and social interaction. Sociology enables us to see how individual behavior is largely shaped by the groups to which we belong and the society in which we live. The sociological imagination helps us to understand how seemingly personal troubles, such as suicide, actually are related to larger social forces including those that are related to global interdependence. Sociology emerged out of the social upheaval produced by industrialization and urbanization in the late eighteenth century. Some early social thinkers including Auguste Comte, Harriet Martineau, Herbert Spencer, and Emile Durkheim emphasized social order and stability; others including Karl Marx, Max Weber, and George Simmel focused on conflict and social change. From its origins in Europe, sociology spread to the United States in the 1890s when the first department of sociology was established at the University of Chicago. Sociologists have traditionally used three primary theoretical perspectives to examine social life: (1) functionalist perspectives assume that society is a stable, orderly system; (2) conflict perspectives assume that society is a continuous power struggle among competing groups, often based on class, race, ethnicity, or gender; and (3) symbolic interactionist perspectives focus on how people make sense of their everyday social interactions. However, a fourth perspective--postmodernism--emerged and gained acceptance among some social thinkers more recently. Sociologists conduct research to gain a more accurate understanding of society. Sociological research is based on an approach that answers questions through a direct, systematic collection and analysis of data. Many sociologists engage in quantitative research, which focuses on data that can be measured numerically. Other research is qualitative, based on interpretive description rather than statistics. Research Models are tailored to the specific problem being investigated, and the focus of the researcher and may be quantitative or qualitative. The following are steps in the conventional quantitative research: (1) select and define the research problem, (2) review previous research, (3) formulate the hypothesis, (4) develop the research design, (5) collect and analyze the data, and (6) draw conclusions and report the findings. Researchers taking the qualitative approach might (1) formulate the problem to be studied instead of creating a hypothesis, (2) collect and analyze the data and (3) report the results. Research methods, systematic techniques for conducting research, include surveys, secondary analysis, field research, and experiments. Many sociologists use multiple methods in order to gain a wider scope of data and points of view. The study of people raises important ethical issues in sociological research. The American Sociological Association sets forth basic standards that sociologists must follow in conducting research.

LEARNING OBJECTIVES

After reading Chapter 1, you should be able to:

1. Describe the sociological imagination and explain its importance in understanding people's behavior.

2. Explain what C. Wright Mills meant by the sociological imagination and why it requires us to include many points of view and diverse experiences in our own thinking.

3. Define social class, gender, and race, and explain why these terms are important to the development of our sociological imaginations.

4. Discuss industrialization and urbanization as factors that contributed to the development of sociological thinking.

5. Identify Auguste Comte, Harriet Martineau, and Herbert Spencer, and explain their unique contributions to early sociology.

6. Contrast Emile Durkheim and Karl Marx's perspectives on society and social conflict.

7. State the major assumptions of functionalism, conflict theory, and interactionism, and identify the major contributors to each perspective.

8. Explain the assumptions of postmodernism.

9. Compare and contrast the views of Max Weber and George Simmel.

10. Differentiate between quantitative and qualitative research and give examples of each.

11. Describe the six steps in the conventional research process, which focuses on deduction and quantitative research.

12. Describe the key steps in conducting qualitative research.

13. Describe the major ethical concerns in sociological research.

KEY TERMS (defined at page number shown and in glossary)

anomie 13
conflict perspectives 17
content analysis 31
control group 33
correlation 33
dependent variable 23
ethnography 33
experiment 33
experimental group 33
functionalist perspectives 15
high-income countries 07
hypothesis 23
independent variable 23
industrialization 09
interview 31
latent functions 17
low-income countries 09
macrolevel analysis 19
manifest functions 17
microlevel analysis 19
middle-income countries 09
participant observation 33
positivism 11
postmodern perspectives 21
qualitative research 23
quantitative research 23
reliability 25
research methods 29
secondary analysis 31
social Darwinism 11
social facts 13
society 05
sociological imagination 05
sociology 05
survey 29
symbolic interactionist perspectives 19
theory 15
urbanization 09
validity 25
variable 23

CHAPTER OUTLINE

I. PUTTING SOCIAL LIFE INTO PERSPECTIVE
   A. **Sociology** is the systematic study of human society and social interaction.
   B. Why Study Sociology?
      1. Sociology helps us see the complex connections between our own lives and the larger, recurring patterns of the society and world in which we live.
         a. A **society** is a large social grouping that shares the same geographical territory and is subject to the same political authority and dominant cultural expectations.

   b. When we examine the world order, we become aware of *global interdependence*- a relationship in which the lives of all people are intertwined closely and any one nation's problems are part of a larger global problem.
   c. Sociological research often reveals the limitations of myths associated with commonsense knowledge that guides ordinary conduct in everyday life.

C. The Sociological Imagination
   1. According to sociologist C. Wright Mills, the **sociological imagination** enables us to distinguish between personal troubles and public issues.
   2. It helps one place seemingly personal troubles into a larger social context, such as the example of suicide.

D. The Importance of a Global Sociological Imagination
   1. A global sociological imagination enables us to distinguish among the world's **high-income, middle-income** and **low-income countries**.
   2.. Developing a sociological imagination requires that we take into account perspectives of people.
      a. People in today's world differ by *race*, a term used by many people to specify groups of people distinguished by physical characteristics such as skin color,
      b. *Ethnicity* which refers to the cultural heritage or identity of a group, based on factors such as language or country of origin,
      c. People also differ by *class,* the relative location of a person or group within a larger society, based on wealth, power, prestige, or other valued resources and by *gender,* the meanings, beliefs, and practices associated with sex differences, referred to as femininity and masculinity.

II. THE DEVELOPMENT OF SOCIOLOGICAL THINKING

A. **Industrialization** - the process by which societies are transformed from dependence on agriculture and handmade products to an emphasis on manufacturing and related industries and **urbanization** - the process by which an increasing proportion of a population lives in cities rather than rural areas contributed to the development of sociological thinking.

B. Some early social thinkers were concerned with social order and stability:
   1. **Auguste Comte** coined the term sociology and stressed the importance of **positivism**, a belief that the world can best be understood through scientific inquiry.
   2. **Harriet Martineau's** most influential work was *Society in America* in which she paid special attention to U.S. diversity based on race, class, and gender.
   3. **Herbert Spencer** used an evolutionary perspective to explain stability and change in societies. Known as a supporter of **Social Darwinism**, he coined the term "survival of the fittest",

equating this process of natural selection with progress and success.

4. According to **Emile Durkheim**, **social facts** are patterned ways of acting, thinking, and feeling that exist outside any one individual and exert social control over each person. **Anomie** is a condition in which social control becomes ineffective as a result of the loss of shared values and of a sense of purpose in society.

C. Other early theorists had differing views on the status quo and stability vs. change:

1. **Karl Marx** believed that conflict especially class conflict -- is inevitable.
   a. Class conflict is the struggle between members of the capitalist class, or *bourgeoisie* and the working class, or *proletariat*.
   b. Exploitation of workers by capitalists results in workers' *alienation*, a feeling of powerlessness and estrangement from other people and from oneself.
2. **Max Weber's** concern with the growth of large-scale organizations is reflected in his work on bureaucracy.
3. **Georg Simmel** emphasized that society is best seen as a web of patterned interactions among people. He assessed the costs of industrialization on individuals.

D. The first U.S. department of sociology was at the University of Chicago.

1. **Robert E. Park** and **George Herbert Mead** were influential early members of the faculty. Park studied urbanization and Mead founded the symbolic interaction perspective.
2. **Jane Addams** wrote *Hull-House Maps and Papers* which was used by other Chicago sociologists for the next forty years.
3. **W.E.B. Du Bois** founded the second U.S. department of sociology at Atlanta University and wrote *The Philadelphia Negro: A Social Study*, examining Philadelphia's African American community.

III. CONTEMPORARY THEORETICAL PERSPECTIVES

A. A theory is a set of logically interrelated statements that attempts to describe, explain, and (occasionally) predict social events. Theories provide a framework or perspective - an overall approach or viewpoint toward some subject for examining various aspects of social life.

B. **Functionalist perspectives** are based on the assumption that society is a stable, orderly system characterized by societal consensus.

1. Societies develop social structures, or institutions, that persist because they play a part in helping society survive. These institutions include: the family, education, government, religion, and economy.
2. **Talcott Parsons** stressed that all societies must make provisions for meeting social needs in order to survive. For example, a division of labor (distinct, specialized functions) between husband and wife is essential for family stability and social order.
3. Robert K. Merton distinguished between intended and unintended functions of social institutions.

a. **Manifest functions** are intended and/or overtly recognized by the participants in a social unit.
b. **Latent functions** are unintended functions that are hidden and remain unacknowledged by participants.
c. *Dysfunctions* are the undesirable consequences of any element of society.

C. According to **conflict perspectives**, groups in society are engaged in a continuous power struggle for control of scarce resources.
1. Along with Karl Marx, Max Weber believed that economic conditions were important in producing inequality and conflict in society; however, Weber also suggested that power and prestige are other sources of inequality.
2. C. Wright Mills believed that the most important decisions in the United States are made largely behind the scenes by the power elite, a small clique composed of the top corporate, political, and military officials.
3. Feminist perspectives focus on patriarchy, a system in which men dominate women and that which is considered masculine is more highly valued than that which is considered feminine.

D. Functionalist and conflict perspectives focus primarily on **macrolevel analysis,** an examination of whole societies, large scale social structures, and social systems. By contrast, symbolic interactionist approaches are based on a **microlevel analysis,** an examination of everyday interactions in small groups rather than large scale social structures.

E. **Symbolic interactionist perspectives** are based on the assumption that society is the sum of the interactions of individuals and groups.
1. George Herbert Mead, a founder of this perspective, emphasized that a key feature distinguishing humans from other animals is the ability to communicate in symbols -- anything that meaningfully represents something else.
2. Some symbolic interactionists focus on people's behavior while others focus on each person's interpretation or definition of a given situation.

F. **Postmodern perspectives** reject existing theories and stress the importance of postindustrialization, consumerism, and global communications in understanding social life.
1. These theories oppose individual academic boundaries, suggesting the sharing of ideas among the various disciplines.
2. The information explosion, the rise of a consumer society, and the emergence of a global village wherein people around the world communicate readily with each other, characterize postmodern societies. This approach opens up broad new avenues of social inquiry.

IV. THE SOCIOLOGICAL RESEARCH PROCESS

A. Research may be either quantitative or qualitative.
1. **Quantitative research** is based on the goal of scientific objectivity and focuses on data that can be measured in numbers.
2. **Qualitative research** uses interpretive description (words) rather than

statistics (numbers) to analyze underlying meanings and patterns of social relationships.

B. The Conventional Research Model
   1. The “conventional” model focuses on quantitative research.
   2. The steps in the conventional research model include:
      a. Selecting and defining the research problem;
      b. Reviewing previous research;
      c. Formulating the hypothesis (if applicable);
      d. Developing the research design;
      e. Collecting and analyzing the data; and
      f. Drawing conclusions and reporting the findings.
   3. Important concepts in the research process:
      a. A *hypothesis* is a statement of the relationship between two or more concepts.
      b. **Variables** are concepts with measurable traits or characteristics that can change or vary from one person, time, situation, or society to another.
      c. The *independent variable* is presumed to cause or determine a dependent variable.
      d. The *dependent variable* is assumed to depend on or be caused by the independent variable(s).
   4. Important concepts in collecting and analyzing data:
      a. A *sample* is the people who are selected from the population to be studied, and should accurately represent that population.
         1. A representative sample is a selection from a larger population that has the essential characteristics of the total population.
         2. A random sample is chosen by chance: every member of an entire population being studied has the same chance of being selected.
      b. **Validity**--the extent to which a study or research instrument accurately measures what it is supposed to measure--and **reliability**--the extent to which a study or research instrument yields consistent results--may be problems in research.
      c. Researchers report their findings in the final stage in order to make the study available for *replication*--the repetition of the investigation in substantially the same way that it was originally conducted.

C. Qualitative research differs from quantitative research in several ways:
   1. Researchers do not always do an extensive literature search before beginning their investigation.
   2. They may engage in problem formulation instead of creating a hypothesis.
   3. This type of research often is built on a collaborative approach in which the "subjects" are active participants in the design process, not just passive objects to be studied.
   4. Researchers tend to gather data in natural settings, such as where the person lives or works, rather than in a laboratory or other research setting.

5. Data collection and analysis frequently occur concurrently, and the analysis draws heavily on the language of the persons studied, not the researcher.

V. RESEARCH METHODS

A. **Research methods** are strategies or techniques for systematically conducting research.

B. **Surveys** are polls in which researchers gather facts or attempt to determine the relationship between facts. Survey data are collected by using self administered questionnaires, personal interviews, and/or telephone surveys.

1. A *questionnaire* is a printed research instrument containing a series of items for the subjects' response. Questionnaires may be self administered by respondents or administered by interviewers in face-to-face encounters or by telephone.
2. An **interview** is a data-collection encounter in which an interviewer asks the respondent questions and records the answers. Survey research often uses structured interviews, in which the interviewer asks questions from a standardized questionnaire.
3. Survey research enables the researcher to do *multivariate analysis*--research involving more than two independent variables.

C. In **secondary analysis of data**, researchers use existing material and analyze data that originally was collected by others.

1. Existing data sources include public records, official reports of organizations or government agencies, and *raw data* collected by other researchers.
2. **Content analysis** is the systematic examination of cultural artifacts or various forms of communication to extract thematic data and draw conclusions about social life.

D. *Field research* is the study of social life in its natural setting: observing and interviewing people where they live, work, and play.

1. In **participant observation**, researchers collect systematic observations while being part of the activities of the groups they are studying.
2. **Ethnography** is a detailed study of the life and activities of a group of people by researchers who may live with that group over a period of years.

E. **Experiments** carefully designed situations in which the researcher studies the impact of certain variables on subjects' attitudes or behavior typically require that subjects be divided into two groups:

1. The *experimental group* contains the subjects who are exposed to an independent variable (the experimental condition) to study its effect on them.
2. The *control group* contains the subjects who are not exposed to the independent variable.
3. The experimental and control groups then are compared to see if they differ in relation to the *dependent variable*, and the *hypothesis* about the relationship of the two variables is confirmed or rejected.

4. Researchers acknowledge that experiments are artificial; social processes that are set up are often not the same as real-life occurrences, for example, the "autocide" research of William Zellner.

VI. ETHICAL ISSUES IN SOCIOLOGICAL RESEARCH

A. The study of people ("human subjects") raises vital questions about ethical concerns in sociological research. Researchers are required to obtain written "informed consent" statements from the persons they study—but what constitutes "informed consent?" And how do researchers protect the identity and confidentiality of their sources?

B. The American Sociological Association (ASA) has a *Code of Ethics* that sets forth certain basic standards sociologists must follow in conducting research. Among these standards are the following:

   1. Researchers must endeavor to maintain objectivity and integrity in their research.
   2. Researchers must safeguard the participants' right to privacy and dignity while protecting them from harm.
   3. Researchers must protect confidential information provided by participants, even when this information is not considered to be "privileged."
   4. Researchers must acknowledge research collaboration and assistance they receive from others and disclose all sources of financial support.

C. Sociologists are committed to adhering to this code and to protecting research participants; however, many ethical issues arise that cannot be resolved easily.

ANALYZING AND UNDERSTANDING THE BOXES

After reading the chapter and studying the outline, re-read the boxes and write down key points and possible questions for class discussion.

Sociology and Everyday Life: How Much Do You Know About Suicide?

Key Points:

Discussion Questions:

1.

2.

3.

Framing Suicide in the Media: Sociology Verses Sensationalism

Key Points:

Discussion Questions:

1.

2.

3.

You Can Make a Difference: Responding to a Cry for Help

Key Points:

Discussion Questions:

1.

2.

3.

PRACTICE TESTS
MULTIPLE CHOICE QUESTIONS
Select the response that best answers the question or completes the statement:

1. Sociology is the systematic study of: (p. 4)
   a. intuition and commonsense knowledge.
   b. human society and social interaction.
   c. the production, distribution, and consumption of goods and services in a society.
   d. personality and human development.

2. All of the following are reasons to study sociology, *except*: (p. 4)
   a. Sociology helps us gain a better understanding of ourselves and our social world.
   b. Sociology utilizes scientific standards to study society.
   c. Sociology confirms the accuracy of common sense knowledge
   d. Sociology helps us look beyond our personal experiences and gain insights into society.

3. According to C. Wright Mills, the sociological imagination refers to the ability to: (p. 5)
   a. distinguish between personal troubles and public issues.
   b. see the relationship between preliterate and literate societies.
   c. be completely objective in examining social life.
   d. seek out one specific cause for a social problem such as suicide.

4. The ability to provide theory and research beyond one's own country enveloping countries all over the world is known as a _____ approach. (p. 6)
   a. global
   b. developed nation
   c. developing nation
   d. personal awareness

5. Femininity and masculinity are _____ - related terms. (p. 8)
   a. sex
   b. gender
   c. biologically
   d. anatomically

6. Sociology is __________ because its practitioners apply both theoretical perspectives and research methods (or orderly approaches) to examinations of social behavior. (p.4)
   a. scientific
   b. methodological
   c. systematic
   d. individualized

7.. Two historical factors that contributed to the development of sociological thinking were: (p.8)
   a. industrialization and urbanization.
   b. industrialization and immigration.
   c. urbanization and centralization.
   d. urbanization and immigration.

8. Each of the following people made an important contribution to early sociology, *except*: (pp. 9-10)
   a. Auguste Comte.
   b. Harriet Martineau.
   c. Adam Smith.
   d. Herbert Spencer.

9. _____ argued that conflict, especially class conflict, is necessary in order to produce social change and a better society. (p. 12)
   a. Auguste Comte
   b. Emile Durkheim
   c. Karl Marx
   d. Harriet Martineau

10. The first U.S. department of sociology was established at _____, where _____ was one of the best known women in the field. (p. 14)
    a. University of Chicago; Jane Addams
    b. Harvard University; Harriet Martineau
    c. University of California; Arlie Hochschild
    d. Princeton University; Sara McLanahan

11. W. E. B. Du Bois referred to the identity conflict of being a black and an American as: (p. 15)
    a. group consciousness.
    b. the American dilemma.
    c. false consciousness.
    d. double-consciousness.

12. _____ perspectives are based on the assumption that society is a stable, orderly system. (p. 15)
    a. Functionalist
    b. Symbolic interactionist
    c. Conflict
    d. Feminist
    e. Postmodernist

13. _____ perspectives are based on the assumption that groups are engaged in a continuous power struggle for control of scarce resources. (p. 17)
    a. Functionalist
    b. Symbolic interactionist
    c. Conflict
    d. Feminist

14. According to your text, all of the following are conflict theorists, *except*: (p. 17)
    a. Max Weber.
    b. Talcott Parsons.
    c. Karl Marx.
    d. C. Wright Mills.

15. The _______ approach directs attention to women's experiences and the importance of gender as an element of social structure. (p. 18)
    a. functionalist
    b. symbolic interactionist
    c. conflict
    d. feminist

16. The _________ perspective examines the rise of a consumer society and the emergence of a global village. (p. 20)
    a. conflict
    b. functionalist
    c. postmodern
    d. symbolic interactionist

17. According to the _____ perspective, society is the sum of the interactions of individuals and groups. (p. 19)
    a. functionalist
    b. symbolic interactionist
    c. conflict
    d. feminist

18. Signs, gestures, written language, and shared values are all examples of: (p. 19)
    a. symbols.
    b. psychological defense mechanisms.
    c. norms.
    d. roles.

19. According to symbolic interactionists, subjective reality is: (p. 19)
    a. innate
    b. acquired
    c. shared
    d. both b and c

20. A study of suicidal behavior based on suicide notes is an example of _____ research. (p. 22)
    a. quantitative.
    b. qualitative.
    c. nonscholarly.
    d. triangulated.

21. The first step in the research process is to: (p. 22)
    a. select and define the research problem.
    b. review previous research.
    c. develop a research design.
    d. formulate the hypothesis.

22. According to Emile Durkheim, a high suicide rate is symptomatic of: (p. 23)
    a. individual problems.
    b. psychological abnormalities.
    c. small scale group problems.
    d. a lack of social integration in society.

23. In Emile Durkheim's study of suicide, the degree of social integration was the: (p. 23)
    a. operational definition.
    b. dependent variable.
    c. independent variable.
    d. spurious correlation.

24. A _____ sample is a selection from a larger population and has the essential characteristics of the total population. (p. 23)
    a. selective
    b. random
    c. representative
    d. longitudinal

25. _____ is the extent to which a study or research instrument accurately measures what it is supposed to measure; _____ is the extent to which a study or research instrument yields consistent results. (p. 23-24)
    a. Validity; replication
    b. Replication; validity
    c. Validity; reliability
    d. Reliability; validity

26. Suppose we are investigating the primary causes of suicide in the late 1990s; upon looking into recent cases of suicide, we find out that a number of the people had just lost their jobs; that they had been unemployed off and on for the past ten years; that they had no religious affiliation; and that a number of them had been divorced within the past five years. This analysis reflects what the text terms: (p. 31)
    a. singular analysis.
    b. multivariate involvement.
    c. plural association.
    d. multivariate analysis.

27. Researchers who use existing material and analyze data that originally was collected by others are engaged in: (p. 31)
    a. unethical conduct.
    b. primary analysis.
    c. secondary analysis.
    d. survey analysis.

28. Observation and ethnography are examples of: (p. 32)
    a. survey research.
    b. experiments.
    c. secondary analysis of existing data.
    d. field research.

29. In an experiment, the subjects in the control group: (p. 33)
    a. are exposed to the independent variable.
    b. are not exposed to the independent variable.
    c. are exposed to the dependent variable.
    d. are not exposed to the dependent variable.

30. According to your text, one important ethical concern of the research of Zellner was that he did not tell his subjects that he was actually researching: (p. 34)
    a. murder.
    b. theft.
    c. autocide.
    d. prostitution.

TRUE-FALSE QUESTIONS

T F 1. Mexico is an example of a low-income country. (p. 8)

T F 2. Industrialization first occurred in France between 1700 and 1850. (p. 8)

T F 3. Auguste Comte is considered by some to be the "founder of sociology." (p. 9-10)

T F 4. In the Marxian framework, class conflict is the struggle between the capitalist class and the working class. (p. 12)

T F 5. Georg Simmel's major concern was protecting the autonomy of large groups. (p. 13)

T F 6. Postmodern societies are characterized by an information explosion. (p. 20)

T F 7. From a functionalist perspective, people sometimes commit suicide because they have lost the ability to dream. (p. 16)

T F 8. According to Robert K. Merton, manifest functions are intended and/or overtly recognized by the participants in a social unit. (p. 15-16)

T F 9. Symbolic interactionist perspectives are based on a macro level analysis of society. (p. 18)

T F 10. Emile Durkheim emphasized that suicide rates provide better explanations for suicide than do individual acts of suicide. (p. 22)

T F 11. With qualitative research, statistics are used to analyze patterns of social relationships. (p. 22)

T F 12. Reliability is when a study gives consistent results to different research over time. (p. 24)

T F 13. Altruistic suicide occurs when individuals are excessively integrated into society. (p. 24)

T F 14. Survey research enables a researcher to do multivariate analysis. (p. 31)

T F 15. Correlation exists when two variables are associated more frequently than could be expected by chance. (p. 33)

SOCIOLOGY IN OUR TIMES: DIVERSITY ISSUES

1. How much do you rely on commonsense knowledge as a guide for your conduct in daily life? What insights do you think you may gain by studying sociology?

2. How are you affected by global interdependence? Can you explain how both personal troubles and public issues are affected by global interdependence?

3. How might a better understanding of the experiences of people who are quite dissimilar from you in regard to race/ethnicity, class, gender, and age help you in your personal life in the future? In your career?

4. Does recent research support the finding that women "attempt" suicide more often than men due to problems in their personal relationships, such as being discarded by a lover or husband? Why is men's suicidal behavior often attributed to performance, such as when their self-esteem and independence are threatened?

5. When you examine your textbooks, do you see pictures of people who are similar to you in gender, race/ethnicity, age, class, and ability/disability? Do you think it is important for young children to see textbook pictures of children similar to themselves? Why or why not?

INTERNET RESOURCES AND EXERCISES: AN INTRODUCTION

The following exercise is intended to help you become familiar with the World Wide Web. By using the Internet, you can access a variety of information systems on a wide range of topics. As

you read the text, note the global perspective utilized throughout the book. The Internet is a major source today of global interdependence, a term defined in your text. The use of the internet has accelerated the process of diffusion. People who have access to and use the internet have now become part of a worldwide cyberculture, a culture right at your fingertips. In order to answer the questions, please follow the instructions to access the web sites given. In order to access a location, type the URL (website location) in the area marked 'location' if you are using Netscape or "address" if you are using Microsoft Explorer.
Search engines (i.e. www.google.com) are a popular means to accessing online information. In Chapters 1-4, you will be provided with suggested engines that can be used to narrow your search to a specific article.

1. Access the Internet and type in your desired search engine's web address. In this case, you will need to type in www.google.com. Next, type in the term counterculture and a list of several articles pertaining to this topic will be pulled up. Select an article and read it to give you a better understanding of countercultures in today's society. Then to have an idea of countercultures that developed in the 1960s, type in www.artsandmusicpa.com in the web address box.
2. Another popular search engine that may assist you in understanding sociological concepts is www.altavista.com . Use Alta Vista to find articles about American urbanization, and the effect urbanization has on public schools. Note: you may have to look through several pages of articles before you are able to come with one that fits your topics. Remember, you can always make your search broader or narrower by eliminating or adding words in the search box. For example, with this specific topic if you added the name of a city, your articles would be limited.

INFOTRAC COLLEGE EDITION ONLINE READINGS AND EXERCISES

In order to do these exercises log on to the Infotrac College Library directly to Infotrac at http://www.infotrac-college.com. Click on Register New Account. You will be taken to the login page where you will need to enter the access code, which came packaged with your textbook. Once you have successfully entered the access code and registered, you will be taken to a screen which will enable you to enter the information to begin your search.
Enter "W.E.B. Du Bois" in the keyword search field, select search in title, citation, abstract. In a few moments a short list will appear which will help you with the following questions. First select the article entitled *Scholars Revisit W.E.B. DuBois, Who Found a New Way to Think About Race in America*. Read the selection and answer the following questions:

a. In "Souls", what did DuBois famously predict?

b. What was significant about DuBois receiving his Ph.D. from Harvard University?

c. In the article, a quote by Mr. Hubbard calls DuBois a founding father of what?

Return to the keyword search list and type in Emile Durkheim. Scroll down to the article entitled *Updating Concepts*. Read the selection and answer the following:

a. Where did Durkheim first use the concept of anomie?

b. Who later adapted the concept of anomie to describe the rift between socially approved societal goals and the means of attaining them?

c. Of the functionalist, conflict, symbolic interactionist, or postmodern approaches, which is the concept of anomie most closely associated with?

STUDENT CLASS PROJECTS AND ACTIVITIES

1. Conduct a secondary analysis of suicide rates in the United States. The U.S. Bureau of the Census and the U.S. Center for Disease Control are some of the possible sources of information. What is the present rate of suicide for all Americans? For females and males? For different races or ethnic groups? For high school students? For young adults? For middle-age adults? For those aged 65 or older? For different regions of the country? Provide any other statistics that assist in depicting suicide rates. Next, select a theoretical perspective that best explains the act of suicide and explain suicide from that perspective, such as, how would that theory describe, explain, and occasionally predict suicide? Submit your paper to your instructor according to any specific instructions.

2. Choose ten specific cartoons from various sources that illustrate some sociological concepts or ideas or any ideas of interest to sociologists as discussed in Chapter 1. Provide a copy of each of the cartoons in your paper. Type at least a paragraph per each cartoon describing what concept or idea or area of sociology the cartoon is illustrating, as well how it illustrates that concept, and why you believe it can be of interest to sociologists. Provide any other additional commentary of each cartoon. Finally, evaluate not only the cartoons, but also the value/worth of the project. Submit your paper to your instructor following any other directions your instructor may have given you in the writing of the paper. Be certain to submit your paper on the correct date!

3. Conduct an informal survey of any 20 people and ask these questions: (1) What is sociology? (2) What do sociologists do? (3) What is your opinion of sociologists? (4) Have you had a course in sociology? If so, where? (5) Have you had a course in a social science? If so, what? (6) Do you know a sociologist? If so, who? (7) Include any other kind of information you may want to obtain from your subjects. (8) Include the names, the gender, the ages, and, if students, the college majors of your subjects. (9) Provide a summary and an evaluation of this project in the writing of this paper. Submit your paper to your instructor following any other directions your instructor may have given you in the writing of this project. (10) Submit your paper on the correct due date.

4. Select a specific historical or contemporary well-known sociological theorist. Your instructor may provide a list for your selection. In this project you are to: (1) investigate the background of the theorist, (2) the life-style of the theorist, and (3) provide any other important bibliographic information on that theorist. (4) Include in your paper the major

works and ideas of your particular theorist and (5) evaluate the contribution that your theorist makes to the field of sociology. (6) Briefly evaluate the value of this project and (7) submit your paper to your instructor on the due date.

ANSWERS TO PRACTICE TESTS FOR CHAPTER 1
Answers to Multiple Choice Questions

1. b Sociology is the systematic study of human society and social interaction. (p. 4)
2. c All of the following are reasons to study sociology, *except*: Sociology confirms the accuracy of common sense knowledge. (p. 4)
3. a According to C. Wright Mills, the sociological imagination refers to the ability to distinguish between personal troubles and public issues. (p. 5)
4. a A global approach provides the researcher an ability to provide theory and research beyond one's own country's approach. (p. 6)
5. b Femininity and masculinity are gender-related terns. (p. 8)
6. a Sociology is systematic in that it includes following set patterns of methods in the correct order. (p.4)
7. a Two historical factors that contributed to the development of sociological thinking were industrialization and urbanization. (p. 8)
8. c Auguste Comte, Harriet Martineau and Herbert Spencer all made important contributions to early sociology. (pp. 9-10)
9. c Karl Marx argued that conflict--especially class conflict--is necessary in order to produce social change and a better society. (p. 12)
10. a The first department of sociology in the United States was established at the University of Chicago where Jane Addams was one of the best known women in the field. (p. 14)
11. d W.E.B. Dubois referred to the identity conflict of being a black and an American as double-consciousness. (p. 15)
12. a Functionalist perspectives are based on the assumption that society is a stable, orderly system. (p. 15)
13. c Conflict perspectives are based on the assumption that groups are engaged in a continuous power struggle for control of scarce resources. (p. 17)
14. b According to your text, all of the following are conflict theorists, <u>except</u> Talcott Parsons. (p. 17)
15. d The feminist approach directs attention to women's experiences and the importance of gender as an element of social structure. (p. 18)
16. c The postmodern perspective examines the rise of a consumer society and the emergence of a global village. (p. 20)
17. b According to the symbolic interactionist perspective, society is the sum of the interactions of individuals and groups. (p. 19)
18. a Signs, gestures, written language, and shared values are all examples of symbols. (p. 19)
19. d Symbolic interactionists believe that subjective reality is acquired and shared through agreed-upon symbols, especially language. (p. 19)

20. b A study of suicidal behavior based on suicide notes is an example of qualitative research. (p. 22)
21. a The first step in the research process is to select and define the research problem. (p. 22)
22. d Durkheim's hypothesis stated that the rate of suicide varies immensely with the degree of social integration. (p. 23)
23. c According to Durkheim's study of suicide, a low degree of social integration was the independent variable. (p. 23)
24. c A representative sample is a selection from a larger population that has the essential characteristics of the total population. (p. 23)
25. c Validity is the extent to which a study or research instrument accurately measures what it is supposed to measure; reliability is the extent to which a study or research instrument yields consistent results. (p. 23-24)
26. d Research involving more than two independent variables enables the researcher to conduct multivariate analysis. (p. 31)
27. c Researchers who use existing material and analyze data that originally was collected by others are engaged in secondary analysis. (p. 31)
28. d Observation and ethnography are examples of field research. (p. 32)
29. b The control group contains the subjects who are not exposed to the independent variable. (p. 33)
30. c William Zellner wanted to examine fatal single-occupant automobile accidents to determine if some drivers were actually committing suicide, but he did not inform his subjects of the real purpose of his research. (p. 34)

Answers to True-False Questions:

1. False Mexico is an example of a middle-income country. (p. 8)

2. False Industrialization first occurred in Britain between 1760 and 1850. (p. 8)

3. True (p. 9-10)

4. True (p. 12)

5. False Simmel's ultimate concern was to protect the autonomy of the individual in society. (p. 13)

6. True (p. 20)

7. True (p. 16)

8. True (p. 15-16)

9. False Symbolic interactionist perspectives are based on a macrolevel analysis of society. (p. 18)

10. True (p. 22)

11. False With qualitative research, interpretive description (words), rather than statistics (numbers) is used to analyze underlying meanings and patterns of social relationships. (p. 22)

12. True (p. 24)

13. True (p. 24)

14. True (p. 31)

15. True (p. 33)

CHAPTER 2
CULTURE

BRIEF CHAPTER OUTLINE

Culture and Society
- Material and Nonmaterial Culture
- Cultural Universals

Components of Culture
- Symbols
- Language
- Values
- Norms

Technology, Cultural Change, and Diversity
- Cultural Change
- Cultural Diversity
- Culture Shock
- Ethnocentrism and Cultural Relativism

A Global Popular Culture?
- High Culture and Popular Culture
- Forms of Popular Culture

Sociological Analysis of Culture
- Functionalist Perspective
- Conflict Perspective
- Symbolic Interactionist Perspective
- Postmodern Perspectives

Culture in the Future

CHAPTER SUMMARY

Culture is the knowledge, language, values, customs, and material objects that are passed from person to person and from one generation to the next. At the macrolevel, culture can be a stabilizing force or a source of discord, conflict, and even violence. At the microlevel, culture is essential for individual survival. Sociologists distinguish between **material culture,** the physical creations of society, and **nonmaterial culture**, the abstract or intangible human creations of society (such as **symbols, language, values,** and **norms**). According to the **Sapir-Whorf hypothesis**, language shapes our perception of reality. For example, language may create and reinforce inaccurate perceptions based on gender, race, ethnicity, or other human attributes. Societies experience cultural change at both material and nonmaterial levels. **Technology**, the knowledge, techniques, and tools that allow people to transform resources into usable form and the knowledge and skills required to use what is developed, continues to shape the material culture. **Cultural lag** -- a gap between the technical development of a society and its moral and legal institutions -- frequently occurs when material culture changes at a faster rate than nonmaterial culture. Cultural change and diversity are intertwined. In the United States, diversity is reflected through race, religion, or other factors. **Subcultures** and **countercultures**

may provide reassurance and social support for people who differ from the dominant group. **Culture shock** refers to the anxiety people experience when they encounter cultures radically different from their own. **Ethnocentrism**--a belief based on the assumption that one's own culture is superior to others, is counterbalanced by **cultural relativism**--the belief that the behaviors and customs of a society must be examined within the context of its own culture. A **global culture**, if it comes into existence, will consist of both **high culture** -- assumed to appeal to primarily the elite -- and **popular culture** -- which is believed to appeal to primarily members of the middle and working class. Functionalist and conflict approaches emphasize the macrolevel workings of society, while the symbolic interactionist approach highlights how people maintain and change culture through their interactions with others. Postmodernist analysis of culture points to the complexity and diversity of the social world and suggests that a new way of conceptualizing culture and society is needed. In the future, the issue of cultural diversity will increase in importance. As technology and global communications continue to have a profound effect on culture, we need to apply our sociological imagination not only to our own society, but the entire world as well.

LEARNING OBJECTIVES

After reading Chapter 2, you should be able to:

1. Explain what culture is and describe how it can be both a stabilizing force and a source of conflict in societies.

2. Describe the importance of culture in determining how people think and act on a daily basis.

3. Describe the importance of language and explain the Sapir-Whorf hypothesis.

4. List and briefly explain ten core values in U.S. society.

5. Contrast ideal and real culture and give examples of each.

6. State the definition of norms and distinguish between folkways, mores, and laws.

7. Distinguish between discovery, invention, and diffusion as means of cultural change. Explain why the rate of cultural change is uneven.

8. Describe subcultures and countercultures; give examples of each.

9. State the definitions for culture shock, ethnocentrism, and cultural relativism, and explain the relationship between these three concepts.

10. Distinguish between high culture and popular culture and between fads and fashion.

11. Describe the functionalist, conflict, and symbolic interactionist perspectives on culture.

12. Explain the postmodernist perspective on culture.

KEY TERMS (defined at page number shown and in glossary)

beliefs 44
counterculture 58
cultural imperialism 61
cultural lag 53
cultural relativism 59
cultural universals 45
culture 39
culture shock 58
ethnocentrism 59
folkways 52
high culture 60
language 47
laws 52
material culture 41
mores 52
non material culture 42
norms 52
popular culture 60
sanctions 52
Sapir-Whorf hypothesis 47
subculture 57
symbol 46
taboos 52
technology 53
value contradictions 51
values 50

CHAPTER OUTLINE

I. CULTURE AND SOCIETY

A. **Culture** is the knowledge, language, values, customs, and material objects that are passed from person to person and from one generation to the next in a human group or society.

B. Culture is essential for the survival of both individuals and societies.

C. As our society becomes more diverse and communication among members of international cultures more frequent, the need to appreciate diversity and to understand how people in other cultures view their world increases.

D. **Material culture** consists of the physical or tangible creations that members of a society make, use, and share while **nonmaterial culture** consists of the abstract or intangible human creations of society that influence people's behavior.

E. According to anthropologist George Murdock, **cultural universals** are customs and practices that occur across all societies. Examples include appearance, activities, social institutions, and customary practices.

II. COMPONENTS OF CULTURE

A. A **symbol** is anything that meaningfully represents something else.

B. **Language** is defined as a set of symbols that expresses ideas and enables people to think and communicate with one another.
   1. According to the **Sapir-Whorf hypothesis**, language not only expresses our thoughts and perceptions but also influences our perception of reality.
   2. Language and gender
      a. Examples of situations in which the English language ignores women include using the masculine gender to refer to human beings in general, and nouns that show the gender of the person we expect in a particular occupation.
      b. Words have positive connotations when relating to male power, prestige, and leadership; when related to women, they convey negative overtones of weakness, inferiority, and immaturity.
   3. Language, race, and ethnicity
      a. Language may create and reinforce our perceptions about race and ethnicity by transmitting preconceived ideas about the superiority of one category of people over another.
      b. The "voice" of verbs may devalue contributions of members of some racial/ethnic groups.

C. **Values** are collective ideas about what is right or wrong, good or bad, and desirable or undesirable in a particular culture.
   1. Ten U.S. core values are:
      a. Individualism;
      b. Achievement and success;
      c. Activity and work;
      d. Science and technology;
      e. Progress and material comfort;
      f. Efficiency and practicality;
      g. Equality;
      h. Morality and humanitarianism;
      i. Freedom and liberty; and
      j. Racism and group superiority.
   2. *Value contradictions* are values that conflict with one another or are mutually exclusive (achieving one makes it difficult to achieve another).
   3. *Ideal culture* refers to the values and standards of behavior that people in a society profess to hold; *real culture* refers to the values and standards of behavior that people actually follow.

D. **Norms** are established rules of behavior or standards of conduct.
   1. **Folkways** are everyday customs that may be violated without serious consequences within a particular culture.

2. **Mores** are strongly held norms that may not be violated without serious consequences within a particular culture. **Taboos** are mores so strong that their violation is considered to be extremely offensive.
3. **Laws** are formal, standardized norms that have been enacted by legislatures and are enforced by formal sanctions.

III. TECHNOLOGY, CULTURAL CHANGE, AND DIVERSITY

A. Societies continually experience cultural change, at both material and nonmaterial levels.
   1. **Technology** refers to the knowledge, techniques, and tools that allow people to transform resources into usable form and the knowledge and skills required to use what is developed.

B. According to William F. Ogburn, **cultural lag** is a gap between the technical development (material culture) of a society and its moral and legal institutions (nonmaterial culture). Cultural lag may result from three processes:
   1. *Discovery* is the process of learning about something previously unknown or unrecognized.
   2. *Invention* is the process of combining existing cultural items into a new form.
   3. *Diffusion* is the transmission of cultural items or social practices from one group or society to another.

C. *Cultural diversity* refers to the wide range of cultural differences found between and within nations.
   1. A **subculture** is a group of people who share a distinctive set of cultural beliefs and behaviors that differ in some significant way from that of the larger society. Examples include Old Order Amish and Chinatowns.
   2. A **counterculture** is a group that strongly rejects dominant societal values and norms and seeks alternative lifestyles. Examples include skinheads and neo-Nazi.

D. **Culture shock** is the disorientation that people feel when they encounter cultures radically different from their own, and they believe they cannot depend on their own taken-for-granted assumptions about life.

E. **Ethnocentrism** is the assumption that one's own culture and way of life are superior to all others. By contrast, **cultural relativism** is the belief that the behaviors and customs of a society must be viewed and analyzed within the context of its own culture.

IV. A GLOBAL POPULAR CULTURE?

A. High culture consists of activities usually patronized by elite audiences while **popular culture** consists of activities, products, and services which are assumed to appeal primarily to members of the middle and working class.

B. *Cultural capital theory* is based on the assumption that high culture is a device used by the dominant class to exclude the subordinate classes.

C. A *fad* is a temporary but widely copied activity followed enthusiastically by large numbers of people.

D. A *fashion* is a currently valued style of behavior, thinking, or appearance that is longer lasting and more widespread than a fad.

E. Will the spread of popular culture produce a homogeneous global culture? Political and religious leaders in some nations oppose this process, which they view as **cultural imperialism** -- the extensive infusion of one nation's culture into other nation's culture.

SOCIOLOGICAL ANALYSIS OF CULTURE

A. According to functionalist theorists, popular culture may be the most widely shared aspect of culture (the "glue") that holds society together.

B. Conflict theorists suggest that cultural values and norms help create and sustain the privileged position of the powerful in society. According to Karl Marx, people are not aware that they are being dominated because they have accepted the *ideology* of society's most powerful members.

C. According to symbolic interactionist theorists, people create, maintain, and modify culture as they go about their everyday activities. People continually negotiate their social realities.

D. Postmodernist theorists believe that much of what has been written about culture in the Western world is Eurocentric and suggest that no single perspective can claim to know social reality; we should deconstruct existing beliefs and theories about culture in hopes of gaining new insights.

VI. CULTURE IN THE FUTURE

A. In future decades, the issue of cultural diversity will increase in importance.

B. Some of the most important changes in cultural patterns may include:
(1) technology,
(2) television and radio, films and videos, and electronic communications will continue to accelerate the flow of information; however,
(3) most of the world's population will not participate in this technological revolution.

C. The study of culture helps us understand not only our own "tool kit" of symbols, stories, rituals, and worldviews, but also to expand our insights to include those of other people of the world who also seek strategies for enhancing their own lives.

ANALYZING AND UNDERSTANDING THE BOXES

After reading the chapter and studying the outline, re-read the four boxes and write down key points and possible questions for class discussion.

Sociology and Everyday Life: How Much Do You Know About Consumption and Credit Cards?

Key Points:

Discussion Questions:

1.

2.

3.

Sociology in a Global Perspective: The Malling of China: What Part Does Culture Play?

Key Points:

Discussion Questions:

1.

2.

3.

You Can Make a Difference: Taking a Stand against Overspending

Key Points:

Discussion Questions

1.

2.

3.

PRACTICE TESTS

MULTIPLE CHOICE QUESTIONS

Select the response that best answers the question or completes the statement:

1. A consumer society is (39):
   a. a capitalist system where citizens participate freely in buying both retail and wholesale.
   b. a system of interrelated spending that account for the stability of the economic system.
   c. a society where discretionary consumption is a mass phenomenon across diverse income categories.
   d. an economic system where patrons of establishments are consumed by the desire to keep buying until they need financial relief.d. an economic system where patrons of establishments are consumed by the desire to keep buying until they need financial relief.

2. _____ consists of knowledge, language, values, customs, and material objects. (p. 39)
   a. Social structure
   b. Society
   c. Culture
   d. Social organization

3. All of the following statements regarding culture are true, *except*: (pp. 39-40)
   a. Culture is essential for our survival.
   b. Culture is essential for our communications with other people.
   c. Culture is fundamental for the survival of societies.
   d. Culture is always a stabilizing force for societies.
   e. Culture consists of both material and nonmaterial aspects.
4. From a sociological perspective, culture: (39-40)
   a. is comprised of people.
   b. seldom generates discord, conflict, or violence.
   c. is interdependent with society.
   d. can exist without society.

5. In discussing the importance of culture, your text points out that: (40)
   a. the members of some societies are born with the information they need to survive.
   b. sharing a common culture with others simplifies day-to-day interactions.
   c. certain societies survive without culture.
   d. culture is viewed the same way by people regardless of their race/ethnicity, class, sex, and age.

6. Spiders' behavior in building webs is an example of a/an: (40)
   a. instinct.
   b. impulse.
   c. reflex.
   d. drive.

7. Humans do not have instincts. Instead, we have: (40-41)
   a. reflexes and drives.
   b. impulses and reflexes.
   c. impulses and drives.
   d. drives and motivations.

8. An unlearned, biologically determined involuntary response to some physical stimuli is known as a/an: (41)
   a. instinct.
   b. reflex.
   c. drive.
   d. learned behavior.

9. Behavioral responses that satisfy needs such as sleep, food, water, or sexual gratification are examples of: (41)
   a. instincts.
   b. reflexes.
   c. drives.
   d. learned behaviors.

10. Which of the following is an example of nonmaterial culture? (p. 42)
   a. The contents of naval shipyards
   b. A redwood forest
   c. A pair of scissors
   d. Language

11. According to the functionalist perspective, cultural universals are: (pp. 44-45)
   a. useful because they ensure the smooth and continued operation of society.
   b. the result of attempts by a dominant group to impose its will on a subordinate group.
   c. independent from functional necessities.
   d. very similar in form from one group to another and from one time to another within the same group.

12. Recent conflicts over use of the Confederate flag have occurred because: (p. 46)
    a. flags are meaningful symbols for some value, belief, or institution and, as such, they can have a wide variety of interpretations.
    b. the United Daughters of the Confederacy wanted to use the flag without acquiring permission from the patent holder.
    c. people generally agree that the flag is not a racist symbol.
    d. some political leaders believe that "flag" discussions are a "small issue" as compared with federal budget debates.

13. _____ can stand for love (a heart on a valentine), peace (a dove), or hate (a Nazi swastika). (p. 46)
    a. symbols
    b. communication
    c. cultural universals
    d. nonmaterial culture

14. The Sapir Whorf hypothesis holds that: (p. 47)
    a. people are imprisoned by their language.
    b. language is solely a human characteristic.
    c. language shapes the view of reality of its speakers.
    d. symbols are more important than language in determining how we view the world.

15. Regarding the relationship between language and gender, the text points out that: (p. 48)
    a. the pronouns he and she are seldom used in everyday conversation by most people.
    b. the English language largely has been purged of sexist connotations.
    c. the English language ignores women by using the masculine form to refer to human beings in general.
    d. words in the English language typically have positive connotations when relating to female power, prestige, and leadership.

16. The most frequently spoken language in U.S. homes other than English is: (p. 50)
    a. Italian.
    b. Spanish.
    c. French.
    d. German.

17. From a _____ perspective, a shared language is essential to a common culture. (p. 62)
    a. functionalist
    b. conflict
    c. feminist
    d. symbolic interactionist

18. Which of the following hypothetical statements does not express a core U.S. value? (pp. 50-51)
    a. "How well does it work?"
    b. "Is this a realistic thing to do?"
    c. "My freedom is important to me."
    d. "It is good to be lazy."

19. All of the following are examples of U.S. folkways, *except*: (p. 52)
    a. using underarm deodorant.
    b. brushing our teeth.
    c. avoiding sexual relationships with siblings.
    d. wearing appropriate clothing for specific occasions.

20. The most common type of formal norms is: (p. 52)
    a. folkways.
    b. mores.
    c. sanctions.
    d. laws.

21. _____ is the process of reshaping existing cultural items into a new form. (p. 54)
    a. Discovery
    b. Diffusion
    c. Invention
    d. Restoration

22. According to the text, all of the following are examples of countercultures, *except*: (pp.57-58)
    a. the Old Order Amish.
    b. drug enthusiasts of the 1970s.
    c. beatniks and flower children.
    d. Freemen of Montana.

23. The disorientation that people feel when they encounter cultures radically different from their own is referred as: (pp. 58-59)
    a. cultural diffusion.
    b. cultural relativism.
    c. cultural disorientation.
    d. cultural shock.

24. The national anthem and the pledge to the flag, may help to foster: (p. 59)
    a. ethnocentrism.
    b. cultural relativism.
    c. cultural diffusion.
    d. cultural indifference.

25. Anthropologist Marvin Harris has pointed out that the Hindu taboo against killing cattle is very important to the economic system in India. This exemplifies: (pp. 59-60)
    a. ethnocentrism.
    b. cultural relativism.
    c. cultural diffusion.
    d. cultural indifference.

26. According to the text, all of the following are examples of high culture, *except*: (p. 60)
    a. live theater.
    b. classical music.
    c. movies.
    d. opera.

27. Sociologist John Lofland divided fads into four major categories; collecting SpongeBob Square Pants trading cards is an example of a/an ________fad. (p. 61)
    a. activity
    b. personality
    c. object
    d. idea

28. According to the text's discussion of popular culture, _________ is the second largest export of the U.S.A. to other nations? (p. 61)
    a. aircraft
    b. popular culture
    c. guns
    d. agricultural produce

29. Conflict theorists point out that ideas are cultural creations of a society's most powerful members who maintain their position through the use of: (pp. 62-63)
    a. ideology.
    b. cultural shock.
    c. cultural relativism.
    d. ethnocentrism.

30. According to the ___ perspective, the focus of culture in the Western world is Eurocentric. (p. 64)
    a. conflict
    b. functionalist
    c. symbolic interaction
    d. postmodern

TRUE-FALSE QUESTIONS

T F 1. Culture is essential for individual survival, as well as, the survival of society. (p. 40)

T F 2. Humans have a number of basic instincts. (p. 40)

T F 3. Material culture consists of creations such as language, values, and rules of behavior. (pp. 41-42)

T F 4. The subject matter of jokes is thought to be a cultural universal. (p. 42-44)

T F 5. As a result of their symbolic significance, flags can be a source of discord and strife among people. (p. 45)

T F 6. According to the Sapir-Whorf hypothesis, language shapes the view of reality of its speakers. (p. 47)

T F 7. Functionalist theorists view language as a source of power and social control that perpetuates inequalities in society. (p. 62)

T F 8. According to sociologist Robin M. Williams, racism and group superiority is a core value in the United States. (p. 51)

T F 9. Unlike folkways, mores have strong moral and ethical connotations that may not be violated without serious consequences. (p. 52)

T F 10. Most technological changes are primarily modifications of existing technology. (p. 53)

T F 11. Cultural lag refers to a gap between the technical development of a society and its moral and legal institutions. (p. 53)

T F 12. Chinatowns and other ethnic subcultures help first-generation immigrants adapt to abrupt cultural change. (pp. 55-57)

T F 13. Fads tend to a long-lasting, widely-spread activity. (pp. 60-61)

T F 14. According to some symbolic interactionist theorists, culture helps people meet their biological, instrumental, and integrative needs. (p. 62)

T F 15. In their examination of culture, postmodernist sociological thinkers make us aware of the fact that no single perspective can group the complexity and diversity of the social world. (p. 64)

SOCIOLOGY IN OUR TIMES: DIVERSITY ISSUES

1. Social analysts suggest that how people view culture is intricately related to their location in society with regard to their race/ethnicity, class, sex, and age. Can you think of examples from your own experiences that either prove or disprove this assertion?

2. Do you think that recent conflicts over the meaning of symbols such as the Confederate flag reflect racism in the United States or are their conflicts "blown out of proportion" by the proponents and opponents in these controversies? What other symbols also may generate controversy?

3. How does the "voice" of verbs in these two sentences influence our perceptions about the achievements or activities of people of color?
   a. "African Americans were given the right to vote."
   b. "Columbus discovered America."

4. What is the relationship between language and gender? What cultural assumptions about women and men does language reflect? Can you provide some examples in which language and gender are intertwined?

INTERNET EXERCISES

You may use either the search engines suggested in Chapter 1 or try the following:

1. Access the Internet and type in your desired search engine's web address. In this case, you will need to type in www.google.com. Next, type in the term cultural lag and a list of several articles pertaining to this topic will be pulled up. Select an article and read it to give you a better understanding of cultural lag in today's society.

2. Use www.excite.com to search for articles concerning ethnocentricity. Read through the articles, select one of your choice and briefly identify examples of ethnocentrism.

INFOTRAC COLLEGE EDITION ONLINE READINGS AND EXERCISES

In order to do these exercises log on to the Infotrac College Library directly to Infotrac at http://www.infotrac-college.com. Click on the login. In the **subject** search field, type in "*Shaping Cultural Tastes at Big Retail Chains*" and click on search. When the list appears, click on "*Shaping Cultural Tastes at Big Retail Chains*" and read the selection. Then answer the following questions:

a. What does the article claim is promoting popular culture?

b. Why are groups like authors, musicians, and civil liberties groups criticizing the chains' power?

Now use your back button to return to the **subject** search page where you typed in "Popular Culture". Click on "The Oprah effect: the $1.4 billion woman influences pop culture, creates stars, and drives entire industries. Here's how she does it." Read the article and answer the following questions:

a. What happened when Lisa Price's product was mentioned on the Oprah Winfrey Show?

b. How does Oprah Winfrey influence popular culture?

c. What are some effects of her influence?

STUDENT CLASS PROJECTS AND EXERCISES

1. Examine the lyrics to ten popular contemporary songs that have to do with identity and the problems of teenagers in America. Country-western, rock, folk, rap, and Christian rock movement are some possible sources. Write your response to the following: (1) Analyze the content of the lyrics in terms of culture; (2) Examine the norms that the lyrics promote; (3) What are the suggestions made by the lyrics? (4) What are the themes of the lyrics? (5) What type of music accompanies the lyrics, and what are some of the impacts of the musical sounds? Write your paper according to the instructions and submit the paper at the appropriate time.

2. Write a paper on this issue: "Are non-industrial cultures an 'endangered species' in the world today?" In this paper, discuss the localizing and globalizing forces of all cultures today. Examine why, in previous obscure cultures, the native cultures appear to be eroding and becoming "acculturated" into the norms of the industrialized world. Is this globalizing force welcomed by the few remaining aboriginal cultures today? Why? Why not ? Is it functional for them? Why? Why not? Select a specific culture where this process may or may not be occurring. For example, you could research the Australian aborigine; the Yanomamo; some of the native groups in Venezuela and other South American cultures; the Mayan Indians of Mexico; the Bushmen in Africa; the Zulu in South Africa; the IK in Uganda, Kenya and the Sudan; the Wajos, Bugis, and Makassarese of Sulawesi, the Navaho or Hopi in the United States, etc. Provide bibliographic references for your data and any other information that you may find relevant to your topic. In your paper, provide a conclusion and a personal evaluation of this project. Be sure to submit your paper on the proper due date!

3. Write a paper on one event that is considered to be the top ten most important cultural events worldwide in the history of humankind. You can choose from one of the top ten of the following list, or you may choose another event, if you prefer. The top ten, according to the staff of TIME MAGAZINE : (1) the Crusades; (2) the signing of the Magna Carta; (3) the Black Death; (4) the Renaissance; (5) the discovery of the Americas; (6) the Reformation; (7) the Industrial Revolution; (8) American and French Revolutions; (9)

World War II; and (10) the Rise and Fall of Communism. Explain why you chose a specific event, and fully describe, discuss, and evaluate that event. You must provide in your paper a historical background to that specific event. You must include in your discussion why this specific event was most influential in worldwide history. In your writing of this project, you must include bibliographic references and an evaluation of this project. Submit your paper according to specifications of your instructor on the assigned due date.

ANSWERS TO PRACTICE TESTS FOR CHAPTER 2

Answers to Multiple Choice Questions

1. c Not simply in any area, but in diverse areas of consumption. (p. 39)
2. c Culture consists of knowledge, language, values, customs, and material objects. (p. 39)
3. d All of the following statements regarding culture are true, *except*: culture is always a stabilizing force for societies. (pp. 39-40)
4. c Everything within society interrelates with everything else in some form. (pp.39-40)
5. b Without a common culture, interaction would be difficult at best. (40)
6. a A difference in humans and other animals has been the ability to override instincts. (p. 40)
7. a Similar to instincts in outcomes but not motivations. (pp. 40-41)
8. b Unlearned without thought. (p. 41)
9. c Unlearned without thought, yet satisfied through thought. (pp.41-42)
10. a Beliefs are not material things. (p.44)
11. a According to the functionalist perspective, cultural universals are useful because they ensure the smooth and continued operation of society. (pp. 44-45)
12. a Symbols such as the flag, especially in times of extreme patriotism can create great conflict and misunderstanding.(p. 46)
13. a Symbols can stand for love (a heart on a valentine), peace, (a dove), or hate (a Nazi swastika). (p.46)
14. c How can one know something except through the language describing it? (p. 47)
15. c Regarding the relationship between language and gender, the text points out that the English language ignores women by using the masculine form to refer to human beings in general. (p. 48)
16. b The most frequently spoken language in U.S. homes other than English is Spanish. (p. 50)
17. a From a functionalist perspective, a shared language is essential to a common culture. (p. 62)
18. d The following hypothetical statement does *not* express a core U.S. value: "It is good to be lazy." (pp. 50-51)
19. c All of the following are examples of U.S folkways, *except*: avoiding sexual relationships with siblings, which is a taboo, not a folkway. (p. 52)
20. d The most common type of formal norms is laws. (p. 52)
21. c Invention is the process of reshaping existing cultural items into a new form. (p. 58)

22. a According to the text, all of the following are examples of countercultures, *except*: the Old Order Amish. (pp. 57-58)
23. d The disorientation that people feel when they encounter cultures radically different from their own is referred to as cultural shock. (pp. 58-59)
24. a The national anthem and the pledge to the flag, may help to foster ethnocentrism. (p. 59)
25. b Anthropologist Marvin Harris has pointed out that the Hindu taboo against killing cattle is very important in the economic system in India. This exemplifies cultural relativism. (pp. 59-60)
26. c According to the text, all of the following are examples of high culture *except*: movies (p. 60)
27. c According to the text, collecting SpongeBob Square Pants trading cards is an example of an object fad. (p. 61)
28. b According to the text's discussion of popular culture, popular culture is the second largest export of the U.S.A. to other nations. (p. 61)
29. a Conflict theorists point out that society's most powerful members maintain their position through the use of ideology. (pp. 62-63)
30. d According to the postmodern perspective, mush of what has been written about culture in the Western world is Eurocentric. (p. 64)

Answers to True-False Questions

1. True (p. 40)
2. False According to sociologists, humans do not have instincts; what we most often think of as instinctive behavior actually can be attributed to reflexes and drives (p. 40)
3. False Nonmaterial culture consists of abstract and intangible human creations such as language, beliefs, values, and rules of behavior. (pp. 41-42)
4. False Although telling jokes may be a universal practice, what is considered to be a joke in one society may be an insult in another. (pp. 42-44)
5. True (p. 45)
6. True (p. 47)
7. False Conflict theorists view language as a source of power and social control; it perpetuates inequalities between people and between groups because words are used to "keep people in their place." (p. 62)
8. True (p. 51)
9. True (p. 52)
10. True (p. 53)
11. True (p. 53)
12. True (pp. 55-57)
13. False Fads tend to be short-lived. (pp. 60-61)
14. False Functionalist theorists, such as anthropologist Bronislaw Malinowski, are the ones who suggest that culture helps people meet their biological, instrumental, and integrative needs. (p. 62)
15. True (p. 64)

CHAPTER 3
SOCIALIZATION

BRIEF CHAPTER OUTLINE

Why Is Socialization Important Around the Globe?
- Human Development: Biology and Society
- Problems Associated with Social Isolation and Maltreatment

Social Psychological Theories of Human Development
- Freud and the Psychoanalytic Perspective
- Piaget and Cognitive Development
- Kolberg and the Stages of Moral Development
- Gilligan's View on Gender and Moral Development

Sociological Theories of Human Development
- Cooley and the Looking-Glass self
- Mead and Role-Thinking
- Recent Symbolic Interactionist Perspectives

Agents of Socialization
- The Family
- The School
- Peer Groups
- Mass Media

Gender and Racial-Ethnic Socialization

Socialization Through the Life Course
- Childhood
- Adolescence
- Adulthood
- Late Adulthood and Ageism

Resocialization
- Voluntary Resocialization
- Involuntary Resocialization

Socialization in the Future

CHAPTER SUMMARY

Socialization is the lifelong process through which individuals acquire a self identity and the physical, mental, and social skills needed for survival in society. Socialization is essential for the individual's survival and for human development; it also is essential for the survival and stability of society. People are a product of two forces: heredity and social environment. Most sociologists agree that while biology dictates our physical makeup, the social environment largely determines how we develop and behave. Humans need social contact to develop properly. Cases of isolated children have shown that people who are isolated during their formative years fail to develop their full emotional and intellectual capacities and that social contact is essential in developing a self. A variety of psychological and sociological theories have been developed to explain child abuse but also to describe how a positive process of socialization occurs. Psychological theories focus primarily on how the individual personality develops. Sigmund Freud developed the psychoanalytic perspective. Jean Piaget pioneered the field of cognitive development, which emphasizes the intellectual development of children. The stages of moral development were developed by Lawrence Kohlberg and later criticized and corrected by Carol Gilligan. Sociological perspectives examine how people develop an awareness of self and learn about their culture. This **self-concept** is not present at birth; it arises in the process of social experience. Charles Horton Cooley developed the image of the **looking-glass self** to explain how people see themselves through the perceptions of others. George Herbert Mead linked the idea of self-concept to **role-taking** and to learning the rules of social interaction. Recent symbolic interactionist perspectives suggest that children are active and creative agents in the socialization process. According to sociologists, **agents of socialization** -- including families, schools, peer groups, and mass media -- teach us what we need to know in order to participate in society. Gender, racial-ethnic, and social class are all determining factors in the life-long socialization process. **Anticipatory socialization**—the process by which knowledge and skills are learned for future roles—often occurs before achieving a new status. **Resocialization** -- the process of learning new attitudes, values, and behaviors, either voluntarily or involuntarily -- sometimes takes place in **total institutions**. In today's global society, because of the rapid pace of technological change, we must learn how to anticipate and consider the consequences of the future.

LEARNING OBJECTIVES

After reading Chapter 3, you should be able to:

1. Define socialization and explain why this process is essential for the individual and society around the globe.

2. Distinguish between sociological and psychological perspectives on the development of human behavior.

3. Explain why cases of isolated children are important to our understanding of the socialization process.

4. Explain Freud's views on the conflict between individual desires and the demands of society.

5. Outline the stages of cognitive development as set forth by Jean Piaget.

6. Compare and contrast the moral development theories of Lawrence Kohlberg and Carol Gilligan.

7. Explain the key components of the theories of Charles Horton Cooley and George Herbert Mead and evaluate the contribution of each to our understanding of the socialization process.

8. Describe Mead's concept of the generalized other and explain why socialization is a two-way process.
9. Explain how a child's positive self-concept develops, according to recent symbolic interactionist perspective.
10. State the major agents of socialization and describe their effects on children's development.
11. Explain what is meant by gender socialization and racial socialization.
12. Outline the stages of the life course and explain how each stage varies based on gender, race, ethnicity, class, and positive or negative treatment.
13. What is ageism, and how does it affect people?
14. Describe the process of resocialization and explain why it often takes place in a total institution.

KEY TERMS (defined at page number shown and in glossary)

ageism 94
agents of socialization 84
anticipatory socialization 91
concrete operational stage 78-79
ego 77
formal operational stage 79
gender socialization 90
generalized other 83
id 77
looking-glass self 81
mass media 87
peer group 87
preoperational stage 78
racial socialization 91
resocialization 94
role-taking 81
self-concept 81
significant other 82
social devaluation 93
socialization 72
sociobiology 73
superego 77
total institution 95

CHAPTER OUTLINE

I. WHY IS SOCIALIZATION IMPORTANT?

   A. **Socialization** is the lifelong process of social interaction through which individuals acquire a self-identity and the physical, mental, and social skills needed for survival in society.
   B. Human Development: Biology and Society

1. Every human being is a product of biology, society, and personal experiences, or heredity and environment.
2. **Sociobiology** is the systematic study of how biology affects social behavior.

C. Problems Associated with Social Isolation and Maltreatment

1. Social environment is a crucial part of an individual's socialization; people need social contact with others in order to develop properly.
2. Researchers have attempted to demonstrate the effects of social isolation on nonhuman primates that are raised without contact with others of their own species.
3. Isolated children illustrate the importance of socialization.
4. The most frequent form of child maltreatment is child neglect.

II. SOCIAL PSYCHOLOGICAL THEORIES OF HUMAN DEVELOPMENT

A. Freud and the Psychoanalytic Perspective

1. In his psychoanalytic perspective, Sigmund Freud divided the mind into three interrelated parts:
   a. The **id** is the component of personality that includes all of the individual's basic biological drives and needs that demand immediate gratification.
   b. The **ego** is the rational, reality-oriented component of personality that imposes restrictions on the innate pleasure-seeking drives of the id.
   c. The **superego** consists of the moral and ethical aspects of personality.
2. When a person is well-adjusted, the ego successfully manages the id and the superego.

B. Piaget and Cognitive Development

1. Jean Piaget's theory of cognitive development is based on the assumption that there are four stages of cognitive development based on how children understand the world around them:
   a. *Sensorimotor Stage* (birth to age 2) children understand the world only though sensory contact and immediate action because they cannot engage in symbolic thought or use language.
   b. *Preoperational Stage* (ages 2-7) children begin to use words as mental symbols and to develop the ability to use mental images.
   c. *Concrete Operational Stage* (ages 7-11) children think in terms of tangible objects and actual events; they also can draw conclusions about the likely physical consequences of an action without always having to try it out.
   d. *Formal Operational Stage* (age 12 through adolescence) adolescents are able to engage in highly abstract thought and understand places, things, and events they have never seen. Beyond this point, changes in thinking are a matter of changes in degree rather than in the nature of their thinking.

C. Kolberg and the Stages of Moral Development

1. Lawrence Kohlberg set forth three levels of moral development:
   a. *Preconventional Level* (ages 7-10) children's perceptions are based on punishment and obedience.

b. *Conventional Level* (age 10 through adulthood) people are most concerned with how they are perceived by their peers and with how one conforms to rules.
c. *Postconventional Level* (few adults reach this stage) people view morality in terms of individual rights. At the final stage of moral development, "moral conduct" is judged by principles based on human rights that transcend government and laws.

2. Some critics have challenged the universality of Kolberg's stage theory.

D. Gillian's view on Gender and Moral Development
1. One of the major critics of Kohlberg's work was psychologist Carol Gilligan, who noted that Kohlberg's model was based solely on male responses.
2. Gilligan believes that men become more concerned with law and order but that women tend to analyze social relationships and the social consequences of behavior.
3. Gillian's argument is that people make moral decisions according to both abstract principles of justice and principles of compassion and care.

III. SOCIOLOGICAL THEORIES OF HUMAN DEVELOPMENT

A. Cooley and the Looking-Glass Self
1. Without social contact, we cannot form a **self-concept** -- the totality of our beliefs and feelings about ourselves.
2. According to Charles Horton Cooley's **looking-glass self**, a person's sense of self is derived from the perceptions of others through a three step process:
   a. We imagine how our personality and appearance will look to other people.
   b. We imagine how other people judge the appearance and personality that we think we present.
   c. We develop a self-concept.
3. Through mutual interrelationships between the individual and society, society shapes people and people shape the society.

B. Mead and Role-Taking
1. George Herbert Mead linked the idea of self-concept to **role-taking** -- the process by which a person mentally assumes the role of another person in order to understand the world from that person's point of view.
   a. **Significant others** are those persons whose care, affection, and approval are especially desired and who are most important in the development of the self; these individuals are extremely important in the socialization process.
   b. Mead divided the self into the "I", the subjective element of the self that represents the spontaneous and unique traits of each person -- and the "me", the objective element of the self, which is composed of the internalized attitudes and demands of other members of society and the individual's awareness of those demands.
   c. Mead outlined three stages of self development:
      (1) *preparatory stage* children largely imitate the people around them;

(2) *play stage* (from about age 3 to 5) children learn to use language and other symbols, thus making it possible for them to pretend to take the roles of specific people;

(3) *game stage* children understand not only their own social position but also the positions of others around them. At this time, the child develops a **generalized other,** an awareness of the demands and expectations of the society as a whole or of the child's subculture.

2. Symbolic interactionist perspectives such as Cooley's and Mead's contribute to our understanding of how the self develops; however, these theories often do not take into account differences in people's experiences based on gender. According to Kasper, Mead's ideas about the social self may be more applicable to men than to women.
3. Recent symbolic interactionist perspectives emphasize that socialization is a collective process in which *children* are active and creative agents, not passive recipients in the socialization process.

IV. AGENTS OF SOCIALIZATION

A. **Agents of socialization** are the persons, groups, or institutions that teach us what we need to know in order to participate in society. These are the most pervasive agents of socialization in childhood:

B. The family is the most important agent of socialization in all societies.

1. Functionalists emphasize that families are the primary focus for the procreation and socialization of children, as well as the primary source of emotional support.
2. To a large extent, the family is where we acquire our specific social position in society.
3. Conflict theorists stress that socialization reproduces the class structure in the next generation.

C. The school has played an increasingly important role in the socialization process as the amount of specialized technical and scientific knowledge has expanded rapidly.

1. Schools teach specific knowledge and skills; they also have a profound effect on a child's self-image, beliefs, and values.
2. From a functionalist perspective, schools are responsible for: (1) socialization and teaching students to be productive members of society; (2) transmission of culture; (3) social control and personal development; and (4) the selection, training, and placement of individuals on different rungs in the society.
3. According to conflict theorists such as Samuel Bowles and Herbert Gintis, much of what happens in school amounts to a hidden curriculum the process by which children from working class and lower income families learn to be neat, to be on time, to be quiet, to wait their turn, and to remain attentive to their work attributes that are important for later roles in the work force.

D. A **peer group** is a group of people who are linked by common interests, equal social position, and (usually) similar age.

1. Peer groups function as agents of socialization by contributing to our sense of "belonging" and our feelings of self worth.
2. Individuals must earn their acceptance with their peers by meeting the group's demands for a high level of conformity to its own norms, attitudes, speech, and dress codes.

E. The mass media is an agent of socialization that has a profound impact on both children and adults.

1. The media function as socializing agents in several ways: (1) they inform us about events; (2) they introduce us to a wide variety of people; (3) they provide an array of viewpoints on current issues; (4) they make us aware of products and services that, if we purchase them, supposedly will help us to be accepted by others; and (5) they entertain us by providing the opportunity to live vicariously (through other people's experiences).

2. Television is the most pervasive form of media; ninety eight percent of all homes in the United States have at least one television set.
3. All mass media, including newspapers, magazines, radio, musical recordings, and books, socialize us in ways that we may or may not be consciously aware of.

V. GENDER AND RACE-ETHNIC SOCIALIZATION

A. **Gender socialization** is the aspect of socialization that contains specific messages and practices concerning the nature of being female or male in a specific group or society.

1. Families, schools, and sports tend to reinforce traditional roles through gender socialization.
2. From an early age, media, including children's books, television programs, movies, and music provide subtle and not so subtle messages about masculine and feminine behavior.

B. **Racial socialization** is the aspect of socialization that contains specific messages and practices concerning the nature of one's racial or ethnic status as it relates to: (1) personal and group identity; (2) intergroup and interindividual relationships; and (3) position in the social hierarchy.

VI. SOCIALIZATION THROUGH THE LIFE COURSE

A. Socialization is a lifelong process: each time we experience a change in status, we learn a new set of rules, roles, and relationships.

1. Even before we enter a new status, we often participate in **anticipatory socialization,** the process by which knowledge and skills are learned for future roles.
2. The most common categories of age are infancy, childhood, adolescence, and adulthood (often subdivided into young adulthood, middle adulthood, and older adulthood).

B. During childhood, family support and guidance are crucial to a child's developing self-concept. In some families children are provided warmth and security; however, some families reflect the discrepancy between cultural ideals and reality where children grow up in a setting characterized by fear, danger, and risks that are created by parental neglect, emotional maltreatment, or premature economic and sexual demands.

C. Anticipatory socialization for adult roles often is associated with adolescence; however, some young people may plunge into adult responsibilities at this time.

D. In early adulthood (usually until about age forty), people work toward their own goals of creating meaningful relationships with others, finding employment, and seeking personal fulfillment. Wilbert Moore divided workplace, or occupational, socialization into four phases:
   1. career choice;
   2. anticipatory socialization;
   3. conditioning and commitment; and
   4. continuous commitment.

E. Between the ages of 40 and 65, people enter middle adulthood, and many begin to compare their accomplishments with their earlier expectations.

F. In older adulthood, some people are quite happy and content; others are not. Difficult changes in adult attitudes and behavior may occur in the last years of life when people experience decreased physical ability and **social devaluation** -- when a person or group is considered to have less social value than other groups.

G. In late adulthood, negative images regarding older persons reinforce **ageism** -- prejudice and discrimination against people on the basis of age, particularly against older persons.

H. It is important to note that everyone does not go through certain passages or stages of a life course at the same age and that life course patterns are strongly influenced by race, ethnicity, and social class, as well.

VII. RESOCIALIZATION

A. **Resocialization** is the process of learning a new and different set of attitudes, values, and behaviors from those in one's previous background and experience.

B. *Voluntary resocialization* occurs when we enter a new status of our own free will (e.g., medical or psychological treatment or religious conversion). *Involuntary resocialization* occurs against a person's wishes and generally takes place within a **total institution** a place where people are isolated from the rest of society for a set period of time and come under the control of the officials who run the institution. Examples include military boot camps, prisons, concentration camps, and some mental hospitals.

VIII. SOCIALIZATION IN THE FUTURE

A. Families are likely to remain the institution that most fundamentally shapes and nurtures personal values and self-identity.

B. However, parents increasingly may feel overburdened by this responsibility, especially without societal support - such as high-quality, affordable child care - and more education in parenting skills.

C. A central issue facing parents and teachers as they socialize children is the growing dominance of the media and other forms of technology.

D. With the rapid pace of technological change, socialization must anticipate and consider the consequences of the future.

ANALYZING AND UNDERSTANDING THE BOXES

After reading the chapter and studying the outline, re-read the boxes and write down key points and possible questions for class discussion.

Sociology and Everyday Life: How Much Do You Know About Early Socialization and Child Care?

Key Points:

Discussion Questions:

1.

2.

3.

Sociology in Global Perspectives: Who Should Pay for Child Care?

Key Points:

Discussion Questions:

1.

2.

3.

You Can Make a Difference: Helping a Child Reach Adulthood

Key Points:

Discussion Questions:

1.

2.

3.

MULTIPLE CHOICE QUESTIONS
Select the response that best answers the question or completes the statement:

1. The example of actress Drew Barrymore's description of childhood maltreatment in her family, is used at the beginning of the chapter on socialization to illustrate that: (pp. 70-71)
   a. large numbers of children experience maltreatment at the hands of family members.
   b. large numbers of children experience maltreatment at the hands of caregivers.
   c. maltreatment is of interest to sociologists because it impacts a child's social growth.
   d. all of the above.

2. _____ is the lifelong process of social interaction through which individuals acquire a self identity. (p. 72)
   a. Human development
   b. Socialization
   c. Behavior modification
   d. Imitation

3. All of the following statements regarding socialization are TRUE, except: (p. 72)
   a. Socialization is essential for the individual's survival and for human development.
   b. Socialization is essential for the survival and stability of society.
   c. Socialization enables a society to "reproduce" itself by passing on cultural content from one generation to the next.
   d. Socialization is a learning process that takes place primarily during childhood.

4. The text points out that "being human": (p. 72)
   a. includes being conscious of ourselves as unique individuals.
   b. is confined to emotions, but does not include ideas.
   c. reflects a fundamental irrationality.
   d. is an inborn characteristic.

5. In the debate over "nature" and "nurture," or heredity and environment, sociologists:
   a. focus on how genetic makeup is a major factor in shaping human behavior. (pp. 72-73)
   b. suggest that genetic inheritance underlies many forms of social behavior, including war and peace.
   c. disagree with the notion that biological principles can be used to explain all human behavior.
   d. do not acknowledge that some aspects of people's physical makeup largely are determined by heredity.

6.. The systematic study of how biology affects social behavior is known as:
   a. socio-physiology. (p. 73)
   b. sociobiology.
   c. sociology.
   d. social psychology.

7. The field of sociobiology was pioneered by: (p. 73)
   a. Harry Harlow and Margaret Harlow
   b. Edward O. Wilson
   c. Sigmund Freud
   d. Wilbert Moore

8. Harry and Margaret Harlow's experiment with rhesus monkeys demonstrated that: (pp. 74-75)
   a. food was more important to the monkeys than warmth, affection, and physical comfort.
   b. monkeys cannot distinguish between a nonliving "mother substitute" and their own mother.
   c. socialization is not important to rhesus monkeys; their behavior is purely instinctive.
   d. without socialization, young monkeys do not learn normal social or emotional behavior.

9. Children whose biological and emotional needs are met view the world: (p. 74)
   a. as a non-trustworthy and abusive place.
   b. with suspicion and fear.
   c. as a safe and comfortable place.
   d. as a hostile place.

10. In discussing child maltreatment, the text points out that: (p. 76)
    a. many types of neglect constitute child maltreatment.
    b. the extent of this problem has been exaggerated by the media.
    c. most sexual abuse perpetrators are punished by imprisonment.
    d. most child maltreatment occurs in families living below the poverty line.

11. Sociologist Kingsley Davis was interested in the case of Anna, a child who was kept in an attic like room, because: (p. 76)
    a. he was studying parental attitudes of young, unmarried mothers.
    b. he wanted to know more about what happens when a child is raised in isolation.
    c. he was attempting to determine how children develop a generalized other.
    d. he was examining the relationship between sexual motives and human behavior.

12. The case of Anna illustrates the effects of ______ abuse. (p. 76)
    a. sexual .
    b. emotional.
    c. physical
    d. all of the above

13. Extensive therapy used as an attempt to socialize Genie: (p. 76)
    a. met with limited success.
    b. met with greater success than sociologists first predicted.
    c. ensured that she could eventually become self-sufficient.
    d. illustrates the effects of socio-biology.

14. Sigmund Freud divided the mind into three interrelated parts; which of the following is NOT one of these? (p. 77)
    a. Id
    b. Ego
    c. Superego
    d. Me

15. According to Sigmund Freud, the _____ consists of the moral and ethical aspects of personality. (p. 77)
    a. id
    b. ego
    c. super ego
    d. libido

16. Which of the following is not one of Jean Piaget's stages of cognitive development? (pp. 78-79)
    a. preoperational
    b. concrete operational
    c. formal operational
    d. post operational

17. In Jean Piaget's theory, during the __________ stage, children think in terms of tangible objects and actual events. (pp. 79-80)
    a. sensorimotor
    b. formal operational
    c. preoperational
    d. concrete operational

18. The stages of moral development were initially set forth by _____ and then criticized by _____. (pp. 78-80)
    a. Charles Horton Cooley George Herbert Mead
    b. Lawrence Kohlberg Carol Gilligan
    c. Jean Piaget Lawrence Kohlberg
    d. Sigmund Freud Jean Piaget

19. All of the following are components of the self concept, *except*: (p. 81)
    a. the physical self.
    b. the active self.
    c. the functional self.
    d. the psychological self.

20. The theories of Charles Horton Cooley and George Herbert Mead can best be classified as _____ perspectives. (pp. 81-84)
    a. symbolic interactionist
    b. functionalist
    c. conflict
    d. feminist

21. According to Charles Horton Cooley, we base our perception of who we are on how we think other people see us and on whether this seems good or bad to us. He referred to this perspective as the: (p. 81)
    a. self-fulfilling prophecy.
    b. generalized other.
    c. looking glass self.
    d. significant other.

22. All of the following are stages in Mead's theory of self development, *except* the:
    a. anticipatory stage. (pp. 82-83)
    b. game stage.
    c. play stage.
    d. preparatory stage

23. The _____ refers to the child's awareness of the demands and expectations of society as a whole or of the child's subculture. (pp. 82-83)
    a. looking glass self
    b. id
    c. ego
    d. generalized other

24. In the __________ stage of self-development, individuals learn to use language and other symbols, thus enabling them to pretend to take the roles of specific people. (p. 82)
    a. preparatory
    b. game
    c. play
    d. generalized

25. During the __________ stage of self-development, individuals understand not only their own social position, but also the position of others around them. (p. 82)
    a. preparatory
    b. play
    c. generalized
    d. game

26. Sheila is playing softball with a group of her friends. Recently, she has been thinking about how her team members need to work as a team when a line drive is hit to center field. In doing this, Sheila is developing an awareness of what George Herbert Mead termed the: (p. 83)
    a. significant other.
    b. generalized other.
    c. looking-glass self.
    d. authoritarian personality.

27. According to the text, the most important agent of socialization in all societies is the: (p. 85)
    a. family.
    b. peer group.
    c. school.
    d. church.

28. Sociologist Melvin Kohn has suggested that _____ is one of the strongest influences on what and how parents teach their children. (p. 85)
    a. race/ethnicity
    b. religion
    c. social class
    d. age

29. Currently, about _____ percent of all U.S. preschool children are in day care of one kind or another. (p. 88)
    a. 15
    b. 25
    c. 60
    d. 75

30. From a conflict perspective, students in the school system: (pp. 86-87)
    a. have different experiences depending on their social class, racial-ethnic background, gender, and other factors.
    b. have blended social experiences due to the equality adjustments built into today's curriculum.
    c. are unrestricted in terms of vertical mobility.
    d. are socialized to be productive members of society.

31. As agents of socialization, peer groups are thought to "pressure" children and adolescents because: (p. 87)
    a. individualism is encouraged and rewarded in these groups.
    b. individuals must earn their acceptance with their peers by conforming to the group's norms.
    c. individuals are encouraged to put friendship above material possessions.
    d. individuals are discouraged from making long-term friends in the peer group.

32. The most important aspects of one's racial identity and attitudes toward other racial-ethnic groups are influenced: (p. 91)
    a. by one's peers
    b. by the media
    c. by one's neighbors
    d. by one's family
    e. by one's college professors

33. The process by which knowledge and skills are learned for future roles is known as: (p. 91)
    a. resocialization.
    b. anticipatory socialization.
    c. cyber socialization.
    d. expectant socialization.
    e. desocialization

34. Social devaluation is most likely to be experienced during this stage of the life course: (p. 92)
    a. infancy and childhood.
    b. adolescence.
    c. early adulthood.
    d. older adulthood.

35. All of the following are examples of voluntary resocialization, *except*: (pp. 94-95)
    a. becoming a student.
    b. going to prison.
    c. becoming a Buddhist.
    d. joining Alcoholics Anonymous.

TRUE FALSE QUESTIONS

T F 1. Around the globe, the process of socialization is essential for the survival of both the society and the individual. (p. 72)

T F 2. In a family in which child abuse occurs, all the children are likely to be victims. (p. 73)

T F 3. Unlike humans, nonhuman primates such as monkeys and chimpanzees do not need social contact with others of their species in order to develop properly. (pp. 74-75)

T F 4. The cases of "Anna" and "Genie" make us aware of the importance of the socialization process because they show the detrimental effects of social isolation and neglect. (p.76)

T F 5. Sigmund Freud theorized that our personalities are largely unconscious hidden away outside our normal awareness. (p. 77)

T F 6. Carol Gilligan's research indicated evidence of male-female differences with regard to morality. (p. 80)

T F 7. According to Charles H. Cooley, our sense of self is permanently fixed. (p. 81)

T F 8. According to George H. Mead, significant others are most important in the development of self. (p. 82)

T F 9. Mead's concept of the generalized other affirms that the self is a social creation. (pp. 83-84)

T F 10. The most pervasive agents of socialization in childhood are the family, the school, and the church. (p. 85)

T F 11. By the time students graduate from high school, they have spent more time in front of the television set than sitting in the classroom. (p. 89)

T F 12. Middle-class families tend to adhere to more rigid gender expectations than do working-class families. (pp. 90-91)

T F 13. According to Patricia Hill Collins, "other-mothers" play a more important part in gender socialization and motivation for African American boys than for girls. (p. 91)

T F 14. In the United States, specific rites of passage exist to mark children's move into adulthood. (p. 92)

T F 15. Voluntary socialization generally takes place within a total institution. (pp. 94-95)

SOCIOLOGY IN OUR TIMES: DIVERSITY ISSUES

1. In what ways have your gender, race, ethnicity, class, and religion shaped your sense of self?
2. When did you first become aware of your racial or ethnic category and of your gender?
3. What effect has racial, ethnic or social class segregation had upon you?
4. Do you think that what we know about human development has been limited by the fact that many theorists have limited their research to white, middle-class respondents? Why or why not?

INTERNET EXERCISES:
You may use the search engines that are listed in the previous chapters; however, the web addresses that are included here will take you directly to articles that will assist you in completing the following exercises.

1. Gender Socialization in American Society

2. Type in the following web page address: www.paho.org/English/AD/GE/VAWMen.pdf. Read the article "Men's Roles in Gender-Based Violence" and answer the following questions. 1. According to the Pan American Health Organization, what is the percentage of Costa Rican women who have reported being beaten while pregnant? 2. According to the fact sheet how many women have reported being the victim of sexual, physical, and psychological assault?

3. Relationship of Ageism to Self-Concept – www.lycos.com
Read the article by Linda M. Woolf, Ph.D. Webster University, and examine the correlation between ageism and self-concept. If for any reason you cannot find the article using Lycos, then try another search engine or go directly to Webster University web page at www.webster.edu.

INFOTRAC COLLEGE EDITION ONLINE READINGS AND EXERCISES

Access the Infotrac website at: http://www.infotrac-college.com in order to answer these questions.
Once you have logged on to Infotrac, type "George Herbert Mead" into the **subject** search field and click return. Click on the periodicals and read the article entitled "Was George Herbert Mead a feminist?"

a. What did Mead strongly oppose and advocate for in 1902 at the University of Chicago?

b. After the reading the article, decide if Mead was a feminist during his time, and if so, would he be considered one today?

Now return to the **subject** search field and type in "Internet Use Said to Cut Into TV Viewing and Socializing." When the result list appears, scroll down to "Internet Use Said to Cut Into TV Viewing and Socializing" and click on it. Read the article to respond to the following.

a. What has the survey found that the internet has displaced?
b. Did the study data answer questions about whether internet use itself strengthened or weakened social relations with one's family and friends?

STUDENT CLASS PROJECTS AND ACTIVITIES

1. Think back to your childhood and list the 4 or 5 televisions programs that you watched most often. Next, list the 4 or 5 programs you watch most often today. Respond in writing to these questions: (1) What values were demonstrated in each of the respected programs-yesteryear and today? (2) Were there, or are there, any stereotypes presented via the programs? If so, what were/are they? (3) Were there/are there any values presented that contradict your parents' values? If so, what were/are they? (4) How do you personally evaluate each of these television programs?

2. Write a five to ten page biography of your life, with special emphasis upon your own socialization processes. Include responses to the following questions: (1) Where and when were you born? What were some of the important societal issues at the time of your birth? Who were/are the major agents of socialization in your life? What were some of your favorite activities as a child? Did you have any favorite toys? If so, what were they? Did they influence your future in any way? If so, how? (2) Who, specifically (by name and status) were the "significant others" in your life? (3) What type of values did your significant others advocate? (4) Who were/are the generalized others in your life? (5) What major social institutions played a major role in your socialization process? How? Why? What impact did they have? (6) In what way did any of the above shape your self-image and goals? How important were your grade-school teachers? (7) In your socialization process, do the theories of George Herbert Mead and Charles Horton Cooley apply to you? If so, explain. (8) Provide any additional information you may want to include in the

writing of your biography. (9) Complete this project by following any specific guidelines provided by your teacher. Evaluate this project and submit your paper at the appropriate time.

ANSWERS TO PRACTICE TESTS FOR CHAPTER 3

Answers to Multiple Choice Questions

1. d — Maltreatment at all levels is common. (pp. 70-71)
2. b — Socialization is the lifelong process of social interaction through which individuals acquire a self-identity. (p. 72)
3. d — It is a lifelong process and does not end at a certain age. (p. 72)
4. a — Being able to separate ourselves from our surroundings, and seeing something distinct. (p. 72)
5. c — The age old question of nature versus nurture. As of yet, no one agrees to how much degree each influences or causes behavior. (pp. 72-73)
6. b — The systematic study of how biology affects social behavior is known as sociobiology. (p. 73)
7. b — He was its originator. (p. 73)
8. d — Harry and Margaret Harlow's experiment with rhesus monkeys demonstrated that without socialization, young monkeys do not learn normal social or emotional behavior. (pp. 74-75)
9. c — It is safe, and in no way do we feel danger. (p. 74)
10. a — In discussing child maltreatment, the text points out that many types of neglect constitute child maltreatment. (p. 76)
11. b — The case of Anna was a classic finding for socialization studies. (p. 76)
12. b — Emotional abuse through isolation. (p.76)
13. a — They are still unsure as to why. (p. 76)
14. d — The “Me” is part of George Herbert Mead’s analysis. (p. 77)
15. c — According to Sigmund Freud, the super ego consists of the moral and ethical aspects of personality. (p. 77)
16. d — The post operational stage is <u>not</u> one of Jean Piaget's stages of cognitive development. (pp. 78-79)
17. d — The separation and classification of events and things begins taking shape. (pp. 79-80)
18. b — The stages of moral development were initially set forth by Lawrence Kohlberg and then criticized by Carol Gilligan. (pp. 78-80)
19. c — The "functional self" is not a component of the self-concept. (p. 81)
20. a — The theories of Charles Horton Cooley and George Herbert Mead can best be classified as symbolic interactionist perspectives. (pp. 81-84)
21. c — According to Charles Horton Cooley, we base our perception of who we are on how we think other people see us and on whether this seems good or bad to us. He referred to this perspective as the looking-glass self. (p. 81)

22. a All of the following are stages in Mead's theory of self development, *except* the anticipatory stage. (pp. 82-83)
23. d The generalized other refers to the child's awareness of the demands and expectations of society as a whole or of the child's subculture. (pp. 82-83)
24. c This overlaps to some degree with the preparatory stage. (p. 82)
25. d At this point the person has a fully developed self and can function very well in society. (82)
26. b A general expectation and understanding of the world around her. (p. 83)
27. a According to the text, the most important agent of socialization in all societies is the family. (p. 85)
28. c Sociologist Melvin Kohn has suggested that social class is one of the strongest influences on what and how parents teach their children. (p. 85)
29. c Currently, about 60 percent of all U.S. preschool children are in day care of one kind or another. (p. 88)
30. a This is true throughout the country. (pp. 86-87)
31. b As agents of socialization, peer groups are thought to "pressure" children and adolescents because individuals must earn their acceptance with their peers by conforming to the group's norms. (p. 87)
32. d The most important aspects of one's racial identity and attitudes toward other racial-ethnic groups are influenced by one's family. (p. 91)
33. b The process by which knowledge and skills are learned for future roles is known as anticipatory socialization. (p. 91)
34. d Social devaluation is most likely to be experienced during older adulthood. (p. 92)
35. b All of the following are examples of voluntary resocialization, *except*: going to prison. (pp. 94-95)

Answers to True-False Questions

1. True (p. 72)
2. False In some families, one child may be the victim of repeated abuse whereas others are not. (p. 73)
3. False Even nonhuman primates need social contact with others of their species in order to develop properly. (pp. 74-75)
4. True (p. 76)
5. True (p. 77)
6. True (p. 80)

7. False According to Cooley, our sense of self is not permanently fixed; it is always developing as we interact with others. (p. 81)

8. True (p. 82)

9. True (pp. 83-84)

10. False The most pervasive agents of socialization in childhood are the family, the school, peer groups, and the mass media. (p. 85)

11. True (p. 89)

12. False Working class families adhere to more rigid gender expectations than do middle-class families. (pp. 90-91)

13. False Sociologists Patricia Hill Collins suggests that "other mothers" play an important part in the gender socialization and motivation of African American children, especially girls. (p. 91)

14. False In the United States, no specific rites of passage exist to mark children's move into adulthood. (p. 92)

15. False <u>Involuntary</u> socialization generally takes place within a total institution. (pp. 94-95)

CHAPTER 4
SOCIAL STRUCTURE AND INTERACTION IN EVERYDAY LIFE

BRIEF CHAPTER OUTLINE

Social Structure: The Macrolevel Perspective
Components of Social Structure
Status
Role
Groups
Social Institutions
Societies: Changes in Social Structure
Durkheim: Mechanical and Organic Solidarity
Tonnies: *Gemeinschaft* and *Gesellschaft*
Industrial and Postindustrial Societies
Social Interaction: The Microlevel Perspective
Social Interaction and Meaning
The Social Construction of Reality
Ethnomethodology
Dramaturgical Analysis
The Sociology of Emotions
Nonverbal Communications
Changing Social Structure and Interaction in the Future

CHAPTER SUMMARY

Social structure and interaction are critical components of everyday life. At the microlevel, **social interaction,** the process by which people act toward or respond to other people, is the foundation of meaningful relationships in society. At the macrolevel, **social structure** is the stable pattern of social relationships that exist within a particular group or society. This structure includes **social institutions**, **groups**, **statuses**, **roles**, and norms. Changes in social structure may dramatically affect individuals and groups, as demonstrated by Durkheim's concepts of **mechanical** and **organic solidarity** and Tonnies' *Gemeinschaft* and *Gesellschaft*. **Industrial societies** are based on technology that mechanizes production. A **postindustrial society** is one in which technology supports a service- and information-based economy. Social interaction within a society is guided by certain shared meanings of how we behave. Race, ethnicity, gender, and social class often influence perceptions of meaning, however. The **social construction of reality** refers to the process by which our perception of reality is shaped by the subjective meaning we give to an experience. **Ethnomethodology** is the study of the commonsense knowledge that people use to understand the situations in which they find themselves. **Dramaturgical analysis** is the study of social interaction that compares everyday life to a theatrical presentation. **Impression management (presentation of self)** refers to efforts to present our own self to others in ways that are most favorable to our own interests or image. Feeling rules shape the appropriate emotions for a given role or specific situation. Social interaction also is marked by **nonverbal communication**, which is the transfer of information between people without the use of speech. The future of the United States may rest upon our collective ability to understand and seek to reduce major social problems at both the macrolevel and microlevel of society.

LEARNING OBJECTIVES

After reading Chapter 4, you should be able to:

1. State the definition of social structure and explain why it is important for individuals and society.
2. State the definition of status and distinguish between ascribed and achieved statuses.
3. Explain what is meant by master status and give at least three examples.
4. Define role expectation, role performance, role conflict, and role strain, and give an example of each.
5. Describe the process of role exiting.
6. Explain the difference in primary and secondary groups.
7. Define formal organization and explain why many contemporary organizations are known as "people processing" organizations.
8. State the definition for social institution and name the major institutions found in contemporary society.
9. Evaluate functionalist and conflict perspectives on the nature and purpose of social institutions.
10. Compare Emile Durkheim's typology of mechanical and organic solidarity with Ferdinand Tonnies' *Gemeinschaft* and *Gesellschaft*.
11. Compare and contrast the characteristics of an industrial society with those of a postindustrial society.
12. Explain what interactionists mean by the social construction of reality.
13. Describe ethnomethodology and note its strengths and weaknesses.
14. Describe Goffman's dramaturgical analysis and explain what he meant by presentation of self.
15. Explain what is meant by the sociology of emotions and describe sociologist Arlie Hochschild's contribution to this area of study.
16. Define nonverbal communication and personal space and explain how these concepts relate to our interactions with others.

KEY TERMS (defined at page number shown and in glossary)

achieved status 105
ascribed status 105
division of labor 114
dramaturgical analysis 122
ethnomethodology 121
face-saving behavior 122
formal organization 112
Gemeinschaft 115
Gesellschaft 115
impression management 122
industrial society 116
master status 105
mechanical solidarity 114
nonverbal communication 124-125
organic solidarity 114
personal space 126
postindustrial society 116
primary group 111

role 108
role conflict 109
role exit 111
role expectation 108
role performance 108
role strain 109-110
secondary group 111
self-fulfilling prophecy 120
social construction of reality 120
social group 111
social institution 112
social interaction 101
social structure 101
status 104
status set 105
status symbols 108

CHAPTER OUTLINE

I. SOCIAL STRUCTURE: THE MACROLEVEL PERSPECTIVE
 A. **Social structure** is the stable pattern of social relationships that exist within a particular group or society.
 B. Social structure creates boundaries that define which persons or groups will be the "insiders" and which will be the "outsiders."
  1. *Social marginality* is the state of being part insider and part outsider in the social structure. Social marginality results in stigmatization.
  2. A *stigma* is any physical or social attribute or sign that so devalues a person's social identity that it disqualifies that person from full social acceptance.

II. COMPONENTS OF SOCIAL STRUCTURE
 A. A **status** is a socially defined position in a group or society characterized by certain expectations, rights, and duties.
  1. A *status set* is comprised of all the statuses that a person occupies at a given time.
  2. Ascribed and achieved statuses:
   a. An **ascribed status** is a social position conferred at birth or received involuntarily later in life. Examples of ascribed statuses include race/ethnicity, age, and gender.
   b. An **achieved status** is a social position a person assumes voluntarily as a result of personal choice, merit, or direct effort. Examples of achieved statuses include occupation, education, and income. Ascribed statuses have a significant influence on the achieved statuses we occupy.
  3. A **master status** is the most important status a person occupies; it dominates all of the individual's other statuses and is the overriding ingredient in determining a person's general social position (e.g., being poor or rich is a master status).
  4. **Status symbols** are material signs that inform others of a person's general social position. Examples include a wedding ring or a Rolls-Royce automobile.

 B. A **role** is a set of behavioral expectations associated with a given status.

1. **Role expectation**--a group's or society's definition of the way a specific role ought to be played--may sharply contrast with **role performance**--how a person actually plays the role.
2. **Role conflict** occurs when incompatible role demands are placed on a person by two or more statuses held at the same time (e.g., a woman whose roles include fulltime employee, mother, wife, caregiver for an elderly parent, and community volunteer).
3. **Role strain** occurs when incompatible demands are built into a single status that a person occupies (e.g., a doctor in a public clinic who is responsible for keeping expenditures down and providing high quality patient care simultaneously). Sexual orientation, age, and occupation frequently are associated with role strain.
4. **Role exit** occurs when people disengage from social roles that have been central to their self-identity (e.g., ex-convicts, ex-nuns, retirees, and divorced women and men).

C. A **social group** consists of two or more people who interact frequently and share a common identity and a feeling of interdependence.

1. A **primary group** is a small, less specialized group in which members engage in face-to-face, emotion-based interactions over an extended period of time (e.g., one's family, close friends, and school or work related peer groups).
2. A **secondary group** is a larger, more specialized group in which the members engage in more impersonal, goal-oriented relationships for a limited period of time (e.g., schools, churches, and corporations).
3. A *social network* is a series of social relationships that link an individual to others.
4. A **formal organization** is a highly structured group formed for the purpose of completing certain tasks or achieving specific goals (e.g., colleges, corporations, and the government).

D. A **social institution** is a set of organized beliefs and rules that establish how a society will attempt to meet its basic social needs. Examples of social institutions include the family, religion, education, the economy, the government, mass media, sports, science and medicine, and the military.

III. SOCIETIES: CHANGES IN SOCIAL STRUCTURE

A. Sociologist Emile Durkheim developed a typology to explain how stability and change occurs in the social structure of societies. He was especially concerned about what happens to social solidarity in a society when a "loss of community" occurs.

B. Tonnies' Typology: Gemeinschaft and Gesellschaft

1. Acccording to Ferdinand Tonnies, the ***gemeinschaft*** is a traditional society in which social relationships are based on personal bonds of friendship and

kinship and on intergenerational stability. Relationships are based on ascribed , rather that achieved statuses. People have a commitment to the entire group and feel a sense of togetherness; however, they have very limited privacy.

2. The ***gesellschaft*** is a large, urban society, in which social bonds are based on impersonal and specialized relationships, with little long-term commitment to the group or consensus on values. Relationships are based on achieved statuses; interactions among people are both rational and calculated. Self-interest dominates, and little consensus exists regarding values

C. Industrial and Postindustrial Societies

1. **Industrial societies** are based on technology that mechanizes production; a large proportion of the population lives in or near cities. Large corporations and government bureaucracies grow in size and complexity
2. **Postindustrial societies** are characterized by service- and information-based economies having information explosions and economies wherein large numbers of people either provide or apply information or are employed in service jobs.

IV. SOCIAL INTERACTION: THE MICROLEVEL PERSPECTIVE

A. Social interaction within a given society has certain shared meanings across situations; however, everyone does not interpret social interaction rituals in the same way. The meanings vary for men and women, whites and people of color, and individuals from different social classes.

B. The **social construction of reality** is the process by which our perception of reality is shaped largely by the subjective meaning that we give to an experience.

C. Our definition of the situation can result in a **self-fulfilling prophecy** a false belief or prediction that produces behavior that makes the original false belief come true.

D. **Ethnomethodology** is the study of the commonsense knowledge that people use to understand the situations in which they find themselves.

1. This approach challenges existing patterns of conventional behavior in order to uncover people's background expectancies, that is, their shared interpretation of objects and events, as well as the actions they take as a result.
2. To uncover people's background expectancies, ethnomethodologists frequently conduct breaching experiments in which they break "rules" or act as though they do not understand some basic rule of social life so that they can observe other people's responses.

E. **Dramaturgical analysis** is the study of social interaction that compares everyday life to a theatrical presentation.

1. This perspective was initiated by Erving Goffman, who suggested that day-to-day interactions have much in common with being on stage or in a dramatic production.
2. Most of us engage in **impression management, or presentation of self,** people's efforts to present themselves to others in ways that are most favorable to their own interests or image.
3. Social interaction, like a theater, has a front stage the area where a player performs a specific role before an audience and a back stage the area where a player is not required to perform a specific role because it is out of view of a given audience.

F. The Sociology of Emotions
   1. Arlie Hochschild suggests that we acquire a set of feeling rules, which shape the appropriate emotions for a given role or specific situation.
   2. Emotional labor occurs when employees are required by their employers to feel and display only certain carefully selected emotions.
   3. Gender, class, and race are related to the expression of emotions necessary to manage one's feelings.

G. **Nonverbal communication** is the transfer of information between persons without the use of speech (e.g. facial expressions, head movements, body positions, and other gestures).

H. **Personal space** is the immediate area surrounding a person that the person claims as private. Age, gender, kind of relationship, and social class are important factors in allocation of personal space. Power differentials between people are reflected in personal space and privacy.

V. CHANGING SOCIAL STRUCTURE AND INTERACTION IN THE FUTURE

A. The social structure in the U.S. has been changing rapidly in the past decades (e.g., more possible statuses for persons to occupy and roles to play than at any other time in history).

B. Ironically, at a time when we have more technological capability, more leisure activities and types of entertainment, and vast quantities of material goods available for consumption, many people experience high levels of stress, fear for their lives because of crime, and face problems such as homelessness.

C. While some individuals and groups continue to show initiative in trying to solve some of our pressing problems, the future of this country rests on our collective ability to deal with major social problems at both the macrolevel (structural) and the microlevel of society.

ANALYZING AND UNDERSTANDING THE BOXES

After reading the chapter and studying the outline, re-read the boxes and write down key points and possible questions for class discussion

Sociology and Everyday Life: How Much Do You Know About Homeless Persons?

Key Points:

Discussion Questions:

1.

2.

3.

Framing Homelessness in the Media: Thematic Framing and Episodic Framing

Key Points:

Discussion Questions:

1.

2.

3.

You Can Make a Difference: One Person's Trash May Be Another Person's Treasure: Recycling for Good Causes

Key Points:

Discussion Questions:

1.

2.

3.

PRACTICE TESTS

MULTIPLE CHOICE QUESTIONS

Select the response that best answers the question or completes the statement:

1. The stable pattern of social relationships that exist within a particular group or society is referred to as: (p. 101)
   a. social structure.
   b. social interaction.
   c. social dynamics.
   d. social constructions of reality.

2. According to the text's discussion of homelessness: (p. 102)
   a. many people choose to be homeless.
   b. most homeless people are mentally ill.
   c. most homeless people are heavy drug users.
   d. one out of four homeless people is a child.

3. About __________ percent of the homeless hold full- or part-time jobs but earn too little to afford housing. (p. 102)
   a. 10
   b. 15
   c. 18
   d. 22

4. The text points out that at the macrolevel, the social structure of a society has several essential elements. Which of the following is NOT one of these? (p. 102)
   a. values
   b. social institutions
   c. groups
   d. roles

5. _________ gives us the ability to interpret the social situations we encounter. (p. 102)
   a. Group solidarity
   b. Human development
   c. Social structure
   d. Social stigma

6. According to the text, newly-arrived immigrants are an example of: (p. 102)
   a. social depletion.
   b. social marginality.
   c. social disintegration.
   d. social misappropriation.

7. A _____ is a socially defined position in a group or society. (p. 104)
   a. location
   b. set
   c. status
   d. role

8. Maxine is an attorney, a mother, a resident of California, and a Jewish American. All of these socially defined positions constitute her: (p. 105)
   a. role pattern.
   b. role set.
   c. ascribed statuses.
   d. status set.

9. A master status: (p. 105)
   a. historically has been held only by men.
   b. is comprised of all of the statuses that a person occupies at a given time.
   c. is the most important status a person occupies.
   d. is a social position a person always assumes voluntarily.

10. Which of the following is NOT an example of ascribed status? (p. 105)
    a. race
    b. ethnicity
    c. gender
    d. education

11. According to the text, a wedding ring and a Rolls-Royce automobile are both examples of: (p. 108)
    a. conspicuous consumption.
    b. status symbols.
    c. status markers.
    d. master status indicators.

12. _____ is (are) the dynamic aspect of a status. (p. 108)
    a. Role
    b. Norms
    c. Groups
    d. People

13. In her study of women athletes in college sports programs, sociologist Tracey Watson found role _____ in the traditionally incongruent identities of being a woman and being an athlete. (p. 109)
    a. strain
    b. distancing
    c. conflict
    d. confusion

14. Role _____ may occur among African American men who have internalized North American cultural norms regarding masculinity yet find it very difficult to attain cultural norms of achievement, success, and power because of racism and economic exploitation. (pp. 109-110)
    a. strain
    b. distancing
    c. conflict
    d. confusion

15. When a homeless person is able to become a domiciled person, sociologists refer to the process as: (p. 111)
    a. role disengagement.
    b. role exit.
    c. role engulfment.
    d. role relinquishment.

16. The text points out that people sometimes *prioritize* their roles, or, on the other hand, they may compartmentalize their lives and "insulate" their various roles. These behaviors are indicative of: (p.109)
    a. role ambiguity.
    b. status frustration.
    c. role performance.
    d. role conflict.

17. According to the text's discussion of role strain: (pp. 108-110)
    a. recent social changes may have decreased role strain in men.
    b. sexual orientation is usually not associated with role strain.
    c. men's traditional position of dominance has eroded as more women have entered the paid work force and demanded more assistance in homemaking responsibilities.
    d. when women acquire their own paycheck, role strain at home often decreases.

18. Role __________ occurs when people consciously foster the impression of a lack of commitment or attachment to a particular role and merely go through the motions of role performance. (p. 110)
    a. distancing
    b. conflict
    c. ambiguity
    d. strain

19. All of the following are examples of a primary group, *except*: (p. 111)
    a. family.
    b. close friends.
    c. peer groups.
    d. students in a lecture hall.

20. Which of the following statements best describes the characteristics of a secondary group? (p. 111)
    a. a small, less specialized group in which members engage in impersonal, goal-oriented relationships over an extended period of time.
    b. a small, less specialized group in which members engage in face-to-face, emotion based interactions over an extended period of time.
    c. a larger, more specialized group in which members engage in face-to-face, emotion based interactions over an extended period of time.
    d. a larger, more specialized group in which members engage in impersonal, goal-oriented relationships over an extended period of time.

21. According to the text, the National Law Center on Homelessness and Poverty and other caregiver groups that provide services for the homeless and others in need are examples of: (p. 112)
    a. primary groups.
    b. formal organizations.
    c. informal organizations.
    d. social networks.

22. Sociologists use the term _____ to refer to a set of organized beliefs and rules that establish how a society will attempt to meet its basic social needs. (p. 112)
    a. social structure
    b. social expectations
    c. social networking
    d. social institutions

23. Emile Durkheim referred to the social cohesion found in industrial societies as: (p. 114)
    a. organic solidarity.
    b. mechanical solidarity.
    c. *Gemeinschaft*.
    d. *Gesellschaft*.

24. Emile Durkheim believed that people in industrial societies come to rely on one another in much the same way that the organs of the human body function interdependently. He referred to this condition as: (pp. 114-115)
    a. mechanical solidarity.
    b. status-oriented bonding.
    c. high-technology bonding.
    d. organic solidarity.

25. Which of the following sociologists used the terms *Gemeinschaft* and *Gesellschaft* to characterize the degree of social solidarity and social control found in societies? (p. 115)
    a. Emile Durkheim
    b. Ferdinand Tonnies
    c. Max Weber
    d. Talcott Parsons

26. The *Gemeinschaft* refers to: (p. 115)
    a. a traditional society in which social relationships are based on personal bonds of friendship and kinship and on intergenerational stability.
    b. a large, urban society in which social bonds are based on impersonal and specialized relationships, with little long-term commitment to the group or consensus on values.
    c. a traditional society in which social relationships have come to be based on corporate procedures.
    d. a postindustrial social order in which traditionalism has been restored.

27. A young person who has been told repeatedly that she or he is not a good student and, as a result, stops studying and receives failing grades is an example of a(n): (p. 120)
    a. selective perception.
    b. objective deduction.
    c. subjective reality.
    d. self-fulfilling prophecy.

28. Ethnomethodologist Harold Garfinkel assigned different activities to his students to see how breaking the unspoken rules of behavior created confusion. His research involved a series of: (pp. 121-122)
    a. shared expectancies.
    b. breaching experiments.
    c. dramaturgical analyses.
    d. impression managements.

29. According to sociologist Arlie Hochschild, people acquire a set of _____, which margins the appropriate emotions for a given role or specific situation. (p. 124)
    a. role expectations
    b. emotional experiences
    c. feeling rules
    d. nonverbal cues

30. According to anthropologist Edward Hall, _____ space is the immediate area surrounding a person that the person claims as private. (p. 126)
    a. private
    b. social
    c. personal
    d. intimate

TRUE-FALSE QUESTIONS

T F 1. Most homeless persons are on the streets by choice or because they were deinstitutionalized by mental hospitals. (p. 102)

T F 2. Sociologists use the term status to refer only to high level positions in society. (p. 104)

T F 3. Historically, the most common master statuses for women have related to positions in the family, such as daughter, wife, and mother. (p. 105)

T F 4. Role performance does not always match role expectation. (pp. 108-109)

T F 5. Role conflict occurs when incompatible role demands are placed on a person by two or more statuses held at the same time. (p. 109)

T F 6. According to the text, married men are more likely than married women to experience role strain in the labor force. (p. 110)

T F 7. Role exit usually does not require a creation of a new identity. (p. 111)

T F 8. Close friends would be an example of a primary group. (p. 111)

T F 9. Social solidarity exists when a group remains cohesive when facing obstacles. (p. 111)

T F 10. Social networks work the same way for men and women and for people from different racial-ethnic groups. (p.112)

T F 11. Symbolic Interactionist theorists argue that social institutions maintain the privileges of the wealthy and powerful while contributing to the powerlessness of others. (p. 114)

T F 12. Conflict theorists emphasize that social institutions exist because they perform essential tasks in a society. (pp. 113-114)

T F 13. Relationships in *gemeinschaft* societies are based on achieved rather than ascribed status. (p. 115)

T F 14. Service economies are more likely found in industrial societies. (p. 116)

T F 15. The need for impression management is most intense when role players have widely divergent or devalued statuses. (p. 122)

SOCIOLOGY IN OUR TIMES: DIVERSITY ISSUES

1. According to the text, "people of color are over represented among the homeless." (p. 110) Using your sociological imagination, how is this "personal trouble" linked to other social problems in our society? If you were to find yourself temporarily homeless, how would you resolve your problem? Would you expect help from a social agency or governmental bureaucracy? Why or why not?

2. The text states that "being poor or rich is a master status that influences many other areas of life, including health, education, and life opportunities." (p. 114) From your own experiences, has your family's social class affected your master status? Do you think your social class has affected your health, education, and life opportunities?

3. Have you experienced role conflict? Role strain? Were your experiences related to your gender, race/ethnicity, class, or age?

4. According to the text, our race/ethnicity, gender, and social class affect the meanings we give to our interaction with others. (p. 127) Have any of these statuses influenced your interpretation of chance encounters while walking down a street?

INTERNET EXERCISES

As mentioned previously, you may use any search engine of your choice; the ones listed here are merely suggestions to assist you in your search.

1. Define status symbol and give five examples of status symbols that exist in present day America. (www.yahoo.com)

2. Read an article concerning the idea of self-fulfilling prophecy. Give a brief paragraph explaining the effect of self-fulfilling prophecy. (www.msn.com)

INFOTRAC COLLEGE EDITION ONLINE READINGS AND EXERCISES

Access the Infotrac web page at **http://www.infotrac-college.com** in order to answer the following questions. When you have entered your password, enter "Body of Evidence: watch what you say—with your eyes, hands, and posture" in the **keyword** search field. When the results list appears, click on the article "Body of Evidence: watch what you say—with your eyes, hands, and posture." Read through the article to help you answer the questions:

a. Mastering what can be a powerful tool?

b. When speaking, besides keeping tabs on your own body language, what else should you do?

Return to the **keyword** search page and enter the "social institution." Scroll down to the article, *A crumbling institution: how social revolution cracked the pillars of marriage."* Read the selection, and answer the following:

a. Give examples in the article of things that have affected the institution of marriage.

b. Why does the article say divorce is becoming a new social institution?

STUDENT CLASS PROJECTS AND ACTIVITIES

1. Utilizing the concept of social networks as discussed in the text, trace your own social networks. Construct diagrams linking yourself to your family and other relatives, and to your friends and to other acquaintances. After constructing the diagrams, answer the following questions/statements: (1) Describe the members in your network (who are they; and what is their gender, approximate age, and current employment, if applicable?). (2) How and why are they linked to you? (3) Summarize the impact of each person in your network; are the ties weak or strong? Why? (4) What persons in your network are most important now? Why? (5) What persons in your network will be especially important to you in future years? (6) Is there any potential for anyone in your network to harm you in any fashion? What can/should you do about this? (7) Is there any potential for anyone in your network to provide you with special advantages for a potential career? Explain in detail your responses to the above questions. Write your paper in the manner described by your instructor and submit your paper on the due date.

2. Social networks can be most useful in securing employment. Conduct an informal survey of 10 people who are fully employed. Provide the basic demographic information about your subjects, such as their age, race, nationality, gender, and address. Ask the following questions: (1) What are their positions? (2) How did they obtain the position, i.e., did personal contacts have anything to do with their securing their specific job? (3) Were these contacts weak or strong ties in the job search? (4) Is the position considered to be a career position? (5) What is the degree of satisfaction of each employee? Record all information and submit your paper to your instructor at the appropriate time and according to the written form provided by your instructor.

3. In the text, the discussion of homelessness provides an avenue for examining the effects of social structure. The examination of that issue and suggested solutions are quite controversial. Some suggest that homelessness is the result of the social system -- there is a shortage of available low-income housing; others suggest that the increase of homelessness is because of serious disabilities and/or dysfunctional behavior of the homeless people. You are to write an issue paper on this topic: "Should more low-income housing be created to help the homeless?" In your paper, (1) provide a background of this problem, responding to questions such as: have any former societal programs providing low-income housing existed? If so, during which presidential administrations? What have been the general attitudes toward low-cost housing today? (2) provide factual information germane to homelessness; (3) provide the positions of opposing viewpoints; (4) provide arguments for and against the issue; (5) draw conclusions; (6) describe your own position and your suggested resolution of homelessness. Note: you may attempt a synthesis of opposing positions, leave the issue essentially unresolved, provide a new position, or different innovative program.

ANSWERS TO PRACTICE TESTS FOR CHAPTER 4

Answers to Multiple Choice Questions

1. a — The stable patterns of social relationships that exist within a particular group or society are referred to as social structure. (p. 101)

2. d — Sad but true. (p. 102)

3. d — The working poor. (p. 102)

4. a — Values play a key role but they are not essential at the macrolevel. (p. 102)

5. c — By being stable, we can assess other things using these as reference points. (p.102)

6. b — According to the text, newly arrived immigrants are examples of social marginality. (p. 102)

7. c — A status is a socially defined position in a group or society. (p. 104)

8. d — Maxine is an attorney, a mother, a resident of California, and a Jewish American. All of these socially defined positions constitute her status set. (p. 105)

9. c — A master status is the most important status a person occupies. (p. 105)

10. d — Access to education can be ascribed but the overall level is achieved. (p. 105)

11. b — According to the text, a wedding ring and a Rolls-Royce automobile are both examples of status symbols. (p. 108)

12. a — Role is the dynamic aspect of a status. (p. 108)

13. c — In her study of women athletes in college sports programs, sociologist Tracey Watson found role conflict in the traditionally incongruent identities of being a woman and being an athlete. (p. 109)

14. a — Role strain may occur among African American men who have internalized North American cultural norms regarding masculinity yet find it very difficult to attain cultural norms of achievement, success, and power because of racism and economic exploitation. (pp. 109-110)

15. b — When a homeless person is able to become a domiciled person, sociologists refer to the process as role exit. (p. 111)

16. d — This can lead to conflict if one role differs with another (p. 109)

17. c A traditional example (pp. 109-110)

18. a We see that with parenting, students, and so-called friends. (p. 110)

19. d All of the following are examples of a primary group, *except*: students in a lecture hall. (p. 111)

20. d The statement that best describes the characteristics of a secondary group is: a larger, more specialized group in which members engage in impersonal, goal-oriented relationships over an extended period of time. (p. 111)

21. b According to the text, the Salvation Army and other caregiver groups that provide services for the homeless and others in need are examples of formal organizations. (p. 112)

22. d Sociologists use the term social institutions to refer to a set of organized beliefs and rules that establish how a society will attempt to meet its basic social needs. (p. 112)

23. a Emile Durkheim referred to the social cohesion found in industrial societies as organic solidarity. (p. 114)

24. d Hence we see the link to organic and the body. (pp. 114-115)

25. b Tonnies, a German sociologist, coined the terms Gemeinschaft and Gesellschaft. (p. 115)

26. a Gemeinschaft is a traditional society in which social relationships are based on personal bonds of friendship and kinship and on intergenerational stability. (p. 115)

27. d A young person who has been told repeatedly that she or he is not a good student and, as a result, stops studying and receives failing grades is an example of self-fulfilling prophecy. (p. 120)

28. b Ethnomethodologist Harold Garfinkel assigned different activities to his students to see how breaking the unspoken rules of behavior created confusion. His research involved a series of breaching experiments. (pp. 121-122)

29. c According to sociologist Arlie Hochschild, people acquire a set of feeling rules, which shape the appropriate emotions for a given role or specific situation. (p. 124)

30. c According to anthropologist Edward Hall, personal space is the immediate area surrounding a person that the person claims as private. (p. 126)

Answers to True-False Questions:

1. False Most homeless persons are not on the streets by choice or because of deinstitutionalization; many hold full- or part-time jobs but earn too little to find an affordable place to live. (p. 102)

2. False In a sociological sense, the term status refers to a socially defined position (regardless of whether it is high-level, mid-level, or low-level) in a group or society characterized by certain expectations, rights, and duties. (p. 104)

3. True (p. 105)

4. True (pp. 106-109)

5. True (p. 109)

6. False Because of assumptions about division of labor and parental responsibilities at home, many married women experience more role strain than married men do. (p. 110)

7. False Role exit involves the creation of a new identity. (p. 111)

8. True (p. 111)

9. True (p. 111)

10. False According to sociologists, social networks work differently for women and men, for different races/ethnicities, and for members of different social classes because of exclusion from "old-boy" social networks. (p.112)

11. False <u>Conflict</u> theorists argue that social institutions maintain the privileges of the wealthy and powerful while contributing to the powerlessness of others. (p. 114)

12. False <u>Functional</u> theorists emphasize that social institutions exist because they perform essential tasks in a society. (pp. 113-114)

13. False Relationships in *gemeinschaft* societies are based on <u>ascribed</u> rather than achieved status. (p. 115)

14. False Service economies are more likely found in postindustrial societies. (p.116)

15. True (p. 122)

CHAPTER 5
GROUPS AND ORGANIZATIONS

BRIEF CHAPTER OUTLINE

Social Groups
    Groups, Aggregates, and Categories
    Types of Groups
Group Characteristics and Dynamics
    Group Size
    Group Leadership
    Group Conformity
    Groupthink
Formal Organizations in Global Perspective
    Types of Formal Organizations
    Bureaucracies
    Problems of Bureaucracies
    Bureaucracy and Oligarchy
Alternative Forms of Organization
    Organizational Structure in Japan
Organizations in the Future

CHAPTER SUMMARY

Groups are a key element of our social structure and much of our social interaction takes place within them. A *social group* is a collection of two or more people who interact frequently, share a sense of belonging, and have a feeling of interdependence. Social groups may be either *primary groups*, small, personal groups in which members engage in emotion based interactions over an extended period, or *secondary groups*, larger, more specialized groups in which members have less personal and more formal, goal oriented relationships. All groups set boundaries to indicate who does and who does not belong: an **ingroup** is a group to which we belong and with which we identify; an **outgroup** is a group we do not belong to or perhaps feel hostile toward. The size of a group is one of its most important features. The smallest groups are **dyads,** groups composed of two members -- and **triads,** groups of three. In order to maintain ties with a group, many members are willing to conform to norms established and reinforced by group members. Three types of *formal organizations,* highly structured secondary groups formed to achieve specific goals in an efficient manner, are normative, coercive, and utilitarian organizations. A **bureaucracy** is a formal organization characterized by hierarchical authority, division of labor, explicit procedures, and impersonality in personnel concerns. The **iron law of oligarchy** refers to the tendency of organizations to become a bureaucracy ruled by the few. A recent movement to humanize bureaucracy has focused on developing human resources. The best organizational structure for the future is one that operates humanely and that includes opportunities for all, regardless of race, gender, or class.

LEARNING OBJECTIVES

After reading Chapter 5 you should be able to:

1. Distinguish between aggregates, categories, and groups from a sociological perspective.
2. Distinguish between primary and secondary groups and explain how people's relationships differ in each.
3. State definitions for ingroup, outgroup, and reference group and describe the significance of these concepts in everyday life.
4. Contrast functionalist and conflict perspectives on the purposes of groups.
5. Describe dyads and triads and explain how interaction patterns change as the size of a group increases.
6. Distinguish between the two functions of leadership and the three major styles of group leadership.
7. Describe the experiments of Solomon Asch and Stanley Milgram and explain their contributions to our understanding about group conformity and obedience to authority.
8. Explain what is meant by groupthink and discuss reasons why it can be dangerous for organizations.
9. Compare normative, coercive, and utilitarian organizations and describe the nature of membership in each.
10. Summarize Max Weber's perspective on rationality and outline his ideal characteristics of bureaucracy.
11. Describe the informal structure in bureaucracies and list its positive and negative aspects.
12. Discuss the major shortcomings of bureaucracies and their effects on workers, clients or customers, and levels of productivity.
13. Describe the iron law of oligarchy and explain why bureaucratic hierarchies and oligarchies go hand in hand.
14. Evaluate U.S. and Japanese models of organization.

KEY TERMS (defined at page number shown and in glossary)

aggregate 134
authoritarian leaders 141
bureaucracy 146
bureaucratic personality 150
category 134
conformity 142
democratic leaders 141
dyad 139
expressive leadership 141
goal displacement 150
groupthink 143
ideal type 146
informal side of bureaucracy 149
ingroup 136
instrumental leadership 141
iron law of oligarchy 152
laissez-faire leaders 141
outgroup 136
rationality 146
reference group 138
small group 139
triad 139

CHAPTER OUTLINE

I. SOCIAL GROUPS
   A. Groups, Aggregates, and Categories
      1. A **social group** is a collection of two or more people who interact frequently with one another, share a sense of belonging, and have a feeling of interdependence.
      2. An **aggregate** is a collection of people who happen to be in the same place at the same time but share little else in common.
      3. A **category** is a number of people who may never have met one another but share a similar characteristic.
   B. Types of Groups
      1. Primary and Secondary Group
         a. According to Charles H. Cooley, a primary group is a small group whose members engage in face-to-face, emotion-based interactions over an extended period of time.
         b. A secondary group is a larger, more specialized group in which the members engage in more impersonal, goal-oriented relationships for a limited period of time.
      2. Ingroups and Outgroups
         a. According to William Graham Sumner, an **ingroup** is a group to which a person belongs and with which the person feels a sense of identity.
         b. An **outgroup** is a group to which a person does not belong and toward which the person may feel a sense of competitiveness or hostility.
      3. Reference Groups
         a. A **reference group** is a group that strongly influences a person's behavior and social attitudes, regardless of whether that individual is an actual member.

b. Reference groups help us explain why our behavior and attitudes sometimes differ from those of our membership groups; we may accept the values and norms of a group with which we identify rather than one to which we belong.

II. GROUP CHARACTERISTICS AND DYNAMICS

A. Group Size

1. A **small group** is a collectively small enough for all members to be acquainted with one another and to interact simultaneously.
2. According to Georg Simmel, small groups have distinctive interaction patterns which do not exist in larger groups.
   a. In a **dyad,** a group composed of two members, the active participation of both members is crucial for the group's survival and members have a more intense bond and a sense of unity not found in most larger groups.
   b. When a third person is added to a dyad, a **triad,** a group composed of three members, is formed, and the nature of the relationship and interaction patterns change.
3. As group size increases, members tend to specialize in different tasks, and communication patterns change.
4. The power relationship depends on both a group's *absolute* size and its *relative size.*

B. Group Leadership

1. Leaders are responsible for directing plans and activities so that the group completes its task or fulfills its goals.
2. Leadership functions:
   a. **Instrumental leadership** is goal or task oriented; if the underlying purpose of a group is to complete a task or reach a particular goal, this type of leadership is most appropriate.
   b. **Expressive leadership** provides emotional support for members; this type of leadership is most appropriate when harmony, solidarity, and high morale are needed.
3. Leadership styles:
   a. **Authoritarian leaders** make all major group decisions and assign tasks to members.
   b. **Democratic leaders** encourage group discussion and decision making through consensus building.
   c. **Laissez-faire leaders** are only minimally involved in decision making and encourage group members to make their own decisions.

C. Group Conformity

1. **Conformity** is the process of maintaining or changing behavior to comply with the norms established by a society, subculture, or other group.
2. In a series of experiments, Solomon Asch found that the pressure toward group conformity was so great that participants were willing to contradict their own best judgment if the rest of the group disagreed with them.

3. Stanley Milgram (a former student of Asch's) conducted a series of controversial experiments and concluded that people's obedience to authority may be more common than most of us would like to believe.
4. Irving Janis coined the term **groupthink** to describe the process by which members of a cohesive group arrive at a decision that many individual members privately believe is unwise.

III. FORMAL ORGANIZATIONS IN GLOBAL PERSPECITVES

A. A formal organization is a highly structured secondary group formed for the purpose of achieving specific goals in the most efficient manner (e.g., corporations, schools, and government agencies).

B. Types of Formal Organizations

1. Amitai Etzioni classified formal organizations into three categories based on the nature of membership.
2. We voluntarily join *normative organizations* when we want to pursue some common interest or to gain personal satisfaction or prestige from being a member. Examples include political parties, religious organizations, and college social clubs.
3. People do not voluntarily become members of *coercive organizations* -- associations people are forced to join. Examples include total institutions, such as boot camps and prisons.
4. We voluntarily join *utilitarian organizations* when they can provide us with a material reward we seek. Examples include colleges and universities, and the workplace.

C. Bureaucracies

1. **Bureaucracy** is an organizational model characterized by a hierarchy of authority, a clear division of labor, explicit rules and procedures, and impersonality in personnel matters.
2. According to **Max Weber**, bureaucracy is the most "rational" and efficient means of attaining organizational goals because it contributes to coordination and control. **Rationality** is the process by which traditional methods of social organization, characterized by informality and spontaneity, gradually are replaced by efficiently administered formal rules and procedures.
3. An **ideal type** is an abstract model which describes the recurring characteristics of some phenomenon; the ideal characteristics of bureaucracy include:
   a. Division of Labor: each member has a specific status with certain assigned tasks to fulfill.
   b. Hierarchy of Authority: a chain of command that is based on each lower office being under the control and supervision of a higher one.
   c. Rules and Regulations: standardized rules and regulations establish authority within an organization and usually are provided to members in a written format.
   d. Qualification-Based Employment: hiring of staff members and professional employees is based on specific qualifications; individual

performance is evaluated against specific standards; and promotions are based on merit as spelled out in personnel policies.

e. Impersonality: interaction is based on status and standardized criteria rather than personal feelings or subjective factors.

4. **The informal side of bureaucracy** is composed of those aspects of participants' day-to-day activities and interactions that ignore, bypass, or do not correspond with the official rules and procedures of the bureaucracy.
    a. The informal structure has also been referred to as *work culture*.
    b. There are two schools of thought about informal structure in organizations; one emphasizes control (or eradication) of informal groups; the other suggests that they should be nurtured.

D. Problems of Bureaucracy
  1. Inefficiency and Rigidity
    a. **Goal displacement** occurs when the rules become an end in themselves (rather than a means to an end), and organizational survival becomes more important than achievement of goals
    b. The term **bureaucratic personality** is used to describe those workers who are more concerned with following correct procedures than they are with getting the job done correctly.
  2. Resistance to Change
  3. Perpetuation of Race, Class, and Gender Inequalities

E. Bureaucracy and Oligarchy
  1. Bureaucracy generates an enormous degree of unregulated and often unperceived social power in the hands of a very few leaders.
  2. According Robert Michels, this results in the **iron law of oligarchy**: the tendency to become a bureaucracy ruled by a few people.

IV. ALTERNATIVE FORMS OF ORGANIZATION

A. "Humanizing" bureaucracy includes: (1) less-rigid, hierarchical structures and greater sharing of power and responsibility; (2) encouragement of participants to share their ideas and try new approaches; and (3) efforts to reduce the number of people in dead-end jobs and to help people meet outside family responsibilities while still receiving equal treatment inside the organization.

B. The Japanese model of organization has been widely praised for its innovative structure, which (until recently) has included:
  1. Lifetime Employment--Workers were guaranteed permanent employment after an initial probationary period.
  2. Quality Circles--small workgroups that meet regularly with managers to discuss the group's performance and working conditions focus on both productivity and worker satisfaction.
  3. Limitations--cultural traditions in Japan emphasize the group while those in the United States emphasize individualism; U.S. workers are less willing to commit to one corporation for their entire work life.

V. ORGANIZATIONS IN THE FUTURE

A. There is a lack of consensus among organizational theorists about the "best" model of organization; however, some have suggested a horizontal model in which both hierarchy and functional or departmental boundaries largely would be eliminated.

B. In the horizontal structure, a limited number of senior executives would still exist in support roles (such as finance and human resources); everyone else would work in multidisciplinary teams that would perform core processes (e.g., product development or sales).

C. It is difficult to determine what the best organizational structure for the future might be; however, everyone can benefit from humane organizational environments that provide opportunities for all people regardless of race, gender, or class.

ANALYZING AND UNDERSTANDING THE BOXES

After reading the chapter and studying the outline, re-read the boxes and write down key points and possible questions for class discussion.

Sociology and Everyday Life: How Much Do You Know About Privacy in Groups and Organizations?

Key Points:

Discussion Questions:

1.

2.

3.

Framing "Community" in the Media

Key Points:

Discussion Questions:

1.

2.

3.

You Can Make a Difference: Creating Small Communities of Our Own Within Large Organizations

Key Points:

Discussion Questions:

1.

2.

3.

PRACTICE TESTS

MULTIPLE CHOICE QUESTIONS

Select the response that best answers the question or completes the statement:

1. Which one of the below requires interacting with others on a face-to-face basis? (p.135)
   a. myspace
   b. facebook
   c. face time
   d. friend space

2. A(n) _____ is a collection of people who happen to be in the same place at the same time while a(n) _____ is a number of people who may never have met one another but share a similar characteristic. (p. 134)
   a. aggregate / category
   b. category / aggregate
   c. social group / aggregate
   d. category / social group

3. People who are the same age, race, and gender, and who share the same educational level constitute a/an: (p. 134)
   a. social group.
   b. aggregate.
   c. formal organization.
   d. category.

4. John had thought about trying out for the football team, but because he is not very athletic, he decides to hang out with the "stoners" at his school. He harbors some feelings of hostility for the football team. For John, the football team represents a(n): (p. 136)
   a. reference group.
   b. secondary group.
   c. ingroup.
   d. outgroup.

5. Sociologist Charles H. Cooley used the term __________ to describe a small, less specialized group in which members engage in face-to-face, emotion-based interactions over an extended period of time. (p. 135)
   a. secondary group
   b. significant others
   c. secondary group
   d. primary group

6. We have primary relationships with other individuals in our primary groups--that is, with our __________, who frequently serve as role models. (p. 135)
   a. personal others
   b. significant others
   c. consanguineal kin
   d. fictive kin

7. A __________ group is a larger, more specialized grouping in which the members engage in more impersonal, goal-oriented relationships for a limited period of time. (p. 135)
   a. secondary
   b. primary
   c. kinship
   d. social

8. Sociologist __________ coined the terms ingroup and outgroup to describe people's feelings toward members of their own and other groups. (p. 136)
   a. Emile Durkheim
   b. Max Weber
   c. William Graham Sumner
   d. Charles H. Cooley

9. The <u>best</u> definition of a reference group is "a group....": (p.138)
   a. that strongly influences members' behavior and social attitudes.
   b. to which a person belongs and with which the person feels a sense of identity.
   c. that strongly influences a person's behavior and social attitudes, regardless of whether that individual is an actual member.
   d. that consists of two or more people who interact frequently and share a common identity.

10. For a person who strongly believes in the value of human rights and equal opportunity, which of the following would NOT constitute a reference group? (p. 138)
    a. the Ku Klux Klan
    b. the American Civil Liberties Union
    c. the National Organization for Women
    d. the Council on Racial Equality

11. A _____ is an alliance created in an attempt to reach a shared objective or goal. (p. 139)
    a. triad
    b. coalition
    c. dyad
    d. reference group

12. _____leadership is goal or task oriented; _____leadership provides emotional support for members. (p. 141)
    a. authoritarian democratic
    b. authoritarian laissez-faire
    c. expressive instrumental
    d. instrumental expressive

13. A group leader at a business related seminar is only minimally involved with the decisions made by the group and encourages members to make their own choices. This illustrates a(n) _____ leader. (p. 141)
    a. expressive
    b. democratic
    c. laissez-faire
    d. authoritarian

14. _________ is the process of maintaining or changing behavior to comply with the norms established by a society, subculture, or other group. (p.142)
    a. Conformity
    b. Anticipatory socialization
    c. Enforcement
    d. Total expectation

15. In one of Solomon Asch's experiments, a subject was asked to compare the length of lines printed on a series of cards without knowing that all other research "subjects" actually were assistants to the researcher. The results of Asch's experiments revealed that: (p. 142)
    a. 85 percent of the subjects routinely chose the correct response regardless of the assistants' opinions.
    b. 33 percent of the subjects routinely chose to conform to the opinion of the assistants by giving the same (incorrect) answer.
    c. 10 percent of the subjects routinely chose to conform to the opinion of the assistants by giving an incorrect answer.
    d. the opinion of the assistants had no influence on the subjects' opinions.

16. According to your text, Solomon Asch's experiment demonstrated that: (p.142)
    a. group pressure encourages independent decision making by participants.
    b. studies of male college students can be generalized to other populations.
    c. people will bow to social pressure in small group settings.
    d. people will not say (or do) things simply to gain the approval of other people.

17. Thorstein Veblen used the term __________ to characterize situations in which workers have become so highly specialized, or have been given such fragmented jobs to do, that they are unable to come up with creative solutions to problems. (p. 150)
    a. trained incapacity
    b. bureaucratic personality
    c. cheerful robot
    d. organization men

18. In reference to the experiments by Stanley Milgram, the text points out that: (p. 143)
    a. obedience to authority may be less common than most people would like to believe.
    b. many of the subjects questioned the ethics of the experiment.
    c. research such as this raises important questions concerning research ethics.
    d. most of the subjects were afraid to conform because the use of electrical current was involved.

19. __________ is the process by which members of a cohesive group arrive at a decision that many individual members privately believe is unwise. (p.143)
    a. Group conformity
    b. Groupthink
    c. Decisional harassment
    d. Situational decision making

20. According to the text, the tragic 2003 explosion of the space shuttle *Columbia* while preparing to land is an example of: (pp. 143-144)
    a. groupthink.
    b. obedience to authority.
    c. compliance.
    d. the iron law of oligarchy.

21. All of the following are types of formal organizations, *except* _____ organizations. (pp. 144-146)
    a. normative
    b. anomic
    c. coercive
    d. utilitarian

22. Membership in _____ organizations is involuntary. (p. 146)
    a. normative
    b. anomic
    c. coercive
    d. utilitarian

23. According to Max Weber, rationality refers to: (p. 146)
    a. the process by which bureaucracy is gradually replaced by alternative types of organization such as quality circles.
    b. the process by which traditional methods of social organization are gradually replaced by bureaucracy.
    c. the level of sanity (or insanity) of people in an organization.
    d. the logic used by organizational leaders in decision making.

24. All of the following are ideal type characteristics of bureaucratic organizations, as specified by Max Weber, *except*: (pp. 146-147)
    a. coercive leadership.
    b. impersonality.
    c. hierarchy of authority.
    d. division of labor.

25. A "grapevine" that spreads information outside of official channels in the workplace is cited in the text as an example of: (p. 149)
    a. informal structures in bureaucracy.
    b. impersonality.
    c. hierarchy of authority.
    d. rules and regulations.

26. The informal structure of bureaucracy which includes the ideology and practices of workers on the job is referred to as: (p. 149)
    a. Parkinson's Law.
    b. the iron law of oligarchy.
    c. the Peter Principle.
    d. the work culture.

27. All of the following are listed in the text as being problems of bureaucracy, *except*: (pp. 150-151)
    a. impersonality.
    b. inefficiency and rigidity.
    c. resistance to change.
    d. perpetuation of race, class, and gender inequalities.

28. According to sociologist Joe R. Feagin's recent research on racial and ethnic inequalities in organizations, many middleclass African Americans today: (p. 151)
    a. are included in informal communications networks at work.
    b. have mentors who take an interest in furthering their careers.
    c. have found that entry into dominant white bureaucratic organizations has brought about integration for them.
    d. have experienced an internal conflict between the ideal of equal opportunity and the prevailing norms of many organizations.

29. According to Robert Michels, all organizations encounter: (p. 152)
    a. the Peter Principle.
    b. the iron law of oligarchy.
    c. Murphy's Law.
    d. groupthink.

30. The post-World War II Japanese organizational structure has all of the following characteristics *except*: (p. 153)
    a. employers have an obligation not to "downsize" by laying off workers.
    b. managers take pay cuts when their companies are financially strapped.
    c. employers have to compete with each other to keep their workers.
    d. workers have a high level of productivity.

TRUE-FALSE QUESTIONS

T F 1. Facebook and other networking websites are good substitutes for face to face interaction. (pp. 132-133)

T F 2. Although formal organizations are secondary groups, they also contain many primary groups within them. (p. 136)

T F 3. According to functionalists, people form groups to meet instrumental and expressive needs. (p. 138)

T F 4. If one member withdraws from a dyad, the group ceases to exist. (p. 139)

T F 5. Laissez-faire leaders encourage group discussion and decision making through consensus building. (p. 141)

T F 6. The research of Solomon Asch demonstrated that few people actually will bow to social pressure in small-group settings. (p. 142)

T F 7. Stanley Milgrim's study provides evidence that obedience to authority is fairly common . (pp. 142-143)

T F 8. Bureaurcratic structures the most common organized form in governments, businesses, and schools. (p. 146)

T F 9. An ideal type is an abstract model that describes the recurring characteristics of some phenomenon. (p. 146)

T F 10. Impersonality in a bureaucracy requires a detached approach toward clients. (p. 147)

T F 11. The formal structure of an organization has been referred to as "bureaucracy's work culture." (p. 149)

T F 12. The theory of a "dual labor market" has been developed to explain how social class distinctions are perpetuated through different types of employment. (p. 151)

T F 13. Political parties, religious organizations, and fraternities are all examples of normative organizations. (pp. 144-145)

T F 14. Until recently, many large Japanese corporations guaranteed their workers permanent employment after an initial probationary period. (p. 153)

T F 15. Informal networks can not serve as a means of communication and cohesion among individuals. (pp. 149-150)

SOCIOLOGY IN OUR TIMES: DIVERSITY ISSUES

1. Can you think of organizations on your campus or in your community where criteria for membership include the ability to pay initiation fees and/or membership dues? To what extent is class a determining factor in gaining membership in these groups? Do gender, race/ethnicity, religion, age, and other factors figure into the membership equation?

2. If you are a student of color, analyze your experiences in organizations to see if you agree with the statement in your text (p.151) that "some bureaucracies perpetuate inequalities of race, class, and gender because this form of organizational structure creates a specific type of work or learning environment." Can you think of a situation where you found the organizational "climate" to be "chilly" because of your race/ethnicity?

   If you are a white student, analyze your experiences in organizations to see if you can think of a situation in which a person of color was treated differently (or "indifferently") because of her/his race/ethnicity.

3. If you are a student from a low-income family, analyze your experiences in organizations in regard to the statement quoted above. Likewise, if you are a student from a middle to upper-income family, see if you can think of a situation in which a low-income person was treated differently from you because of lack of economic means.

4. If you are a woman, analyze your experiences in organizations in regard to the statement quoted above. Likewise, if you are a man, see if you can think of a situation in which a woman was treated differently from you in an organization because she was a woman.

INTERNET EXERCISES

Use an Internet search engine to aid you in completing the following exercises:

1. Explain the concept of goal displacement.
2. Define groupthink and identify a situation in which groupthink occurs.

INFOTRAC COLLEGE EDITION ONLINE READINGS AND EXERCISES

Access the Infotrac website and log on. Once online, type "Max Weber and Bureaucracy" in the **keyword** search field. When the result list appears, select "Max Weber (1864-1920): the conceptualisation of bureaucracy," and answer these questions after reading the article.

a. How did Weber's principal contribution to the study of organizations originate?

b. What type of power did Weber consider inherent in any hierarchy or bureaucracy?

c. What were some of the shortcomings of bureaucracy?

Now return to the **keyword** search field and type "groupthink" in the search field. Read the selection titled "All Together Now" to answer the following.

a. Give examples of problems from groupthink.

b. What does the author say groupthink is a precondition for?

STUDENT CLASS PROJECTS AND EXERCISES

1. Prior to taking office, the U.S. President-elect announces to the public his/her choice of all the cabinet and chief advisory appointments. These appointments profoundly shape many of the policies and directions the U.S. federal government, which is one of the largest bureaucracies in the country. As a required written project, select ten of these positions. Describe the position, in terms of the purpose of the office; the requirements, if any, for the position; and the length of the term of that position. Name and briefly describe the credentials of the person who occupies that office. Some examples of cabinet posts are: U.S. Attorney General, Agriculture Secretary, Secretary of the Interior, Transportation Secretary, Secretary of State, Secretary of Defense, Education Secretary, Energy Secretary, Secretary of Veterans Affairs, Commerce Secretary, Labor Secretary, Secretary of the Treasury, etc. Some examples of non-cabinet appointments are: Office of Management and Budget, White House Chief of Staff, White House Council of Economic Advisers, Environmental Protection Agency Administrator, etc. Submit your paper at the designated time according to the format and instructions provided by your instructor.

2. One example of a formal organization is your college or university. Gather information about the characteristics of the organization. Is it constructed according to the specific characteristics of a bureaucracy as described by Max Weber? Explain your response. You may gather information from the college catalog, available organizational charts, and/or via interviews with personnel who work in the organization. Draw an organizational chart to demonstrate the characteristics.

3. Visit a factory with an assembly line, preferably a large factory. Have them obtain or construct an organizational chart of the factory. Try to speak with several employees who occupy different positions in the factory. Carefully record their responses to these questions: (1) What is their specific job description? (2) Why do they work there? (3) What is their specific responsibility? (4) Are they satisfied with their position? Why? Why not? (5) What about the formal and informal aspect of the work environment? What is the degree of friendship with other workers or management? (6) Do they have any control over their responsibilities? Explain. (7) Are there any improvements they would make? If so, what specifically? Submit your papers at the appropriate date.

ANSWERS TO PRACTICE TESTS FOR CHAPTER 5

Answers to Multiple Choice Questions

1. c — Face time requires face-to-face interaction with others. (p. 134)
2. a — An aggregate is a collection of people who happen to be in the same place at the same time while a category is a number of people who may never have met one another but share a similar characteristic. (p. 134)
3. d — The brain naturally makes categories so a method of analysis would be to categorize the various social groups. (p. 134)

4. d John had thought about trying out for the football team, but because he is not very athletic, he decides to hang out with the "stoners" at his school. He harbors some feelings of hostility for the football team. For John, the football team represents an outgroup. (p. 136)
5. d Very strong emotional relationships here. (p. 135)
6. b Significant others and primary groups synonymous with each other. (p. 135)
7. a More impersonal is the key phrase. (p. 135)
8. c Sumner was an early sociologist who had a great influence on sociology and it still carries over with sociological language. (p. 136)
9. c A reference group strongly influences a person's behavior and social attitudes. (p.138)
10. a The KKK typically would not be associated with fighting for human rights and equal opportunity. (p. 138)
11. b A coalition is an alliance created in an attempt to reach a shared objective or goal. (p. 139)
12. d Instrumental leadership is goal or task oriented; expressive leadership provides emotional support for members. (p. 141)
13. c A group leader at a business-related seminar is only minimally involved with the decisions made by the group and encourages members to make their own choices. This illustrates a laissez-faire leader. (p. 141)
14. a This is needed for social organization. (p. 142)
15. b In one of Solomon Asch's experiments, a subject was asked to compare the length of lines printed on a series of cards without knowing that all other research "subjects" actually were assistants to the researcher. The results of Asch's experiments revealed that 33 percent of the subjects routinely chose to conform to the opinion of the assistants by giving the same (incorrect) answer. (p. 142)
16. c Something that is now a mainstay in sociological research. (p. 142)
17. a This is due to their inability to see the whole picture. (p. 150)
18. c In reference to the experiments by Stanley Milgram, the text points out that research such as this raises important questions concerning research ethics. . 143)
19. b Associated with conformity and social organization. (p. 143)
20. a According to the text, the tragic 2003 explosion of the space shuttle *Columbia* while preparing to launch is an example of groupthink. (pp. 143-144)
21. b All of the following are types of formal organizations, *except*: anomic organizations. (pp. 144-146)
22. c Membership in coercive organizations is involuntary. (p. 146)
23. b According to Max Weber, rationality refers to the process by which traditional methods of social organization are gradually replaced by bureaucracy. (p. 146)
24. a All of the following are ideal-type characteristics of bureaucratic organizations, as specified by Max Weber, *except*: coercive leadership (pp. 146-147)
25. a A "grapevine" that spreads information outside of official channels in the workplace is cited in the text as an example of informal structures in bureaucracy. (p. 149)

26. d The information structure of a bureaucracy which includes the ideology and practices of workers on the job is referred to as work culture. (p. 149)
27. a All of the following are listed in the text as being shortcomings of bureaucracy, *except*: impersonality. (pp. 150-151)
28. d According to sociologist Joe R. Feagin's research, many middle-class African Americans who enter white bureaucratic organizations today have experienced an internal conflict between the ideal of equal opportunity and the prevailing norms of many organizations. (p. 151)
29. b According to Robert Michels, all organizations encounter the iron law of oligarchy. (p. 152)
30. c According to the organizational structure in Japan, employers do not have to compete with one another to keep workers. (p. 153)

Answers to True-False Questions:

1. False Facebook and other networking websites are good substitutes for face to face interaction. pp. 132-133)

2. True (p. 136)

3. True (p. 138)

4. True (p. 139)

5. False Democratic leaders encourage group discussion and decision making through consensus building; laissez-faire leaders are only minimally involved in decision making and encourage group members to make their own decisions. (p. 141)

6. False Solomon Asch's research indicated that people will bow to social pressure in small-group settings. (p. 142)

7. True (pp. 142-143)

8. True (p. 146)

9. True (p. 146)

10. True (p. 147)

11. False The informal structure of an organization has been referred to as "bureaucracy's work culture," because it includes the ideology and practices of workers on the job. (p. 149)

12. True (p. 151)

13. True (pp. 144-145)

14. True (p. 153)

15. False Informal networks serve as a means of communication and cohesion among individuals. (pp. 149-150)

CHAPTER 6
DEVIANCE AND CRIME

BRIEF CHAPTER OUTLINE

What Is Deviance?

Who Defines Deviance?
What is Social Control?

Functionalist Perspectives on Deviance

What Causes Deviance, and Why Is It Functional for Society?
Strain Theory: Goals and Means to Achieve Them
Opportunity Theory: Access to Illegitimate Opportunities

Conflict Perspectives on Deviance

Deviance and Power Relations
Deviance and Capitalism
Feminist Approaches
Approaches Focusing on Race, Class, and Gender

Symbolic Interactionist Perspectives on Deviance

Differential Association Theory and Differential Reinforcement Theory
Control Theory: Social Bonding
Labeling Theory

Postmodernist Perspectives on Deviance

Crime Classifications and Statistics

How the Law Classifies Crime
Other Crime Categories
Crime Statistics
Terrorism and Crime
Street Crimes and Criminals
Crime Victims

The Criminal Justice System

The Police
The Courts
Punishment and Corrections

Deviance and Crime in the United States in the Future

The Global Criminal Economy

CHAPTER SUMMARY

By definition, **deviance** is any behavior, belief, or condition that violates significant social norms in the society or group in which it occurs. **Crime** is a form of deviant behavior that violates criminal law and is punishable by fines, jail terms, and other sanctions. A subcategory, **juvenile delinquency,** refers to a violation of a law or the commission of a status offence by young people. All societies create norms in order to define, reinforce, and to help teach acceptable behavior. They also have various mechanisms of **social control,** systematic practices developed by social groups to encourage conformity and to discourage deviance. What causes deviance, and why is it functional for society? Functionalists suggest that deviance is inevitable in all societies and serves several functions: it clarifies rules, unites groups, and promotes social change. Functionalists use **strain theory** and opportunity theory, such as **access to illegitimate opportunity structures** and to argue that socialization into the core value of material success without the corresponding legitimate means to achieve that goal accounts for much of the crime committed by people from lower-income backgrounds, especially when a person's ties to society are weakened or broken. **Conflict theorists** suggest that people with economic and political power define as criminal any behavior that threatens their own interests and are able to use the law to protect their own interests. **Symbolic Interactionists** use **differential association theory**, **differential reinforcement theory, social bonding theory** and **labeling theory** to explain how a person's behavior is influenced and reinforced by others. Liberal, radical and socialist feminist theorists provide alternative approaches to explain deviance and crime. Postmodernists' perspectives on deviance examine the intertwining nature of knowledge, power, and technology on social control and discipline. The law classifies crime into felonies and misdemeanors based on the seriousness of the crime: **Uniform Crime Report** (UCR) is the major source of information on crimes reported in the United States. General categories of crime include: **violent crime, property crime, public order crime**, **occupational** and **corporate crime**, **organized crime**, and **political crime**. Studies show that many more crimes are committed than are reported in official crime statistics. In explaining terrorism and crime, the rational choice approach suggests that terrorists are rational actors who constantly calculate the gains and losses of participation in violent—and often suicidal—acts against others. Gender, age, social class, and race are key factors in official street crime statistics. The criminal justice system includes the police, the courts, and punishment; these agencies often have considerable discretion in dealing with deviance. As we move into the future, we need new approaches for dealing with crime and delinquency. Equal justice under the law needs to be guaranteed, regardless of race, class, gender, or age. Global crime -- the networking of powerful criminal organizations and their associates in shared activities around the world – has expanded rapidly in the era of global communications and rapid transportation networks. Reducing global crime will require a global response, including the cooperation of law enforcement agencies around the world.

LEARNING OBJECTIVES

After reading Chapter 6, you should be able to:

1. Explain the nature of deviance and describe its most common forms.

2. Discuss the functions of deviance from a functionalist perspective and outline the principal features of strain, opportunity, and control theories.

3. Describe the key components of differential association theory and differential reinforcement theory from the symbolic interactionist perspective.

4. Identify rational choice, social bonding, and labeling theories and explain why they are examples of interactionist perspectives.

5. Discuss conflict perspectives on deviance and note the strengths and weaknesses of critical and feminist approaches to deviance and crime.

6. Examine the simultaneous effects of race, class, and gender on deviant behavior.

7. Define and explain the postmodernist perspectives on deviance.

8. Explain the legal classification of crime.

9. Describe the types of behavior included in conventional crime categories.

10. Differentiate between occupational and corporate crime and explain "criminals."

11. Describe organized crime and political crime and explain how each may weaken social control in a society.

12. Explain why official crime statistics may not be an accurate reflection of the actual number and kinds of crimes committed in the United States.

13. Explain how the rational choice approach explains terrorism and crime.

14. Describe the criminal justice system and explain how police, courts, and prisons have considerable discretion in dealing with offenders.

15. State the four functions of punishment and explain how disparate treatment of the poor, all people of color, and white women is evident in the U.S. prison system.

16. Compare and contrast possible patterns of deviance and crime in the United States in the future.

17. Explain the characteristics of the global criminal economy.

KEY TERMS (defined at page number shown and in glossary)

corporate crime 175
criminal justice system 181
crime 162
criminology 163
deviance 160
differential association theory 169
illegitimate opportunity structures 165
juvenile delinquency 162
labeling theory 170
occupational (white-collar) crime 175
organized crime 176
political crime 176
primary deviance 171
property crimes 174
punishment 184
secondary deviance 171
social bond theory 170
social control 162
strain theory 164
terrorism 178
tertiary deviance 171
victimless crimes 175
violent crimes 173

CHAPTER OUTLINE

I. WHAT IS DEVIANCE?

A. By definition, **deviance** is any behavior, belief or condition that violates significant social norms in the society or group in which it occurs.

1. Deviance is relative and it varies in its degree of seriousness: some forms of deviant behavior are officially defined as a crime.
2. **Crime** is a form of deviant behavior that violates criminal law and is punishable by fines, jail terms, and other negative sanctions.
3. **Juvenile delinquency** refers to a violation of law or the commission of a status offence by young people.

B. What is social control?

1. All societies have norms that govern acceptable behavior and mechanisms of **social control**, systematic practices developed by social groups to encourage conformity and discourage deviance.
2. **Criminology** is the systematic study of crime and the criminal justice system, including police, courts, and prisons.

II. FUNCTIONALIST PERSPECTIVES ON DEVIANCE

A. What causes deviance, and why is it functional for society?

1. Emile Durkheim regarded deviance as a natural and inevitable part of all societies.
2. Deviance is universal because it serves three important functions:
   a. Deviance clarifies rules.
   b. Deviance unites a group.
   c. Deviance promotes social change.
3. Functionalists acknowledge that deviance also may be dysfunctional for society; if too many people violate the norms, everyday existence may become unpredictable, chaotic, and even violent.

B. Strain Theory: Goals and Means to Achieve Them

1. According to **strain theory**, people feel strain when they are exposed to cultural goals that they are unable to obtain because they do not have access to culturally approved means of achieving those goals. Robert Merton identified five ways in which people adapt to cultural goals and approved ways of achieving them:
   a. Conformity;
   b. Innovation;
   c. Ritualism;
   d. Retreatism; and
   e. Rebellion.

C. Opportunity Theory: Access to Illegitimate Opportunities

1. According to Richard Cloward and Lloyd Ohlin, for deviance to occur people must have access to **illegitimate opportunity structures** –

circumstances that provide an opportunity for people to acquire through illegitimate activities what they cannot achieve through legitimate channels.

III. CONFLICT PERSPECTIVES ON DEVIANCE

A. Deviance and Power Relations

1. According to conflict theorists, people in positions of power maintain their advantage by using the law to protect their own interests.

B. Deviance and Capitalism

1. According to the critical approach, the way laws are made and enforced benefits the capitalist class by ensuring that individuals at the bottom of the social class structure do not infringe on the property or threaten the safety of those at the top.

C. Feminist Approaches

1. While there is no single feminist perspective on deviance and crime, three schools of thought have emerged:

a. Liberal feminism is based on the assumption that women's deviance and crime is a rational response to gender discrimination experienced in work, marriage, and interpersonal relationships.

b. Radical feminism is based on the assumption that women's deviance and crime is related to patriarchy (male domination over females) that keeps women more tied to family, sexuality, and home, even if women also have full-time paid employment.

c. Socialist feminism is based on the assumption that women's deviance and crime is the result of women's exploitation by capitalism and patriarchy (e.g., their overrepresentation in relatively low-wage jobs and their lack of economic resources).

2. Feminist scholars of color have pointed out that these schools of feminist thought do not include race and ethnicity in their analyses.

D. Approaches Focusing on Race, Class, and Gender

1. Some recent studies have focused on the simultaneous effects of race, class, and gender on deviant behavior.

2. Feminist sociologists and criminologists believe that research on women as both victims and perpetrators of crime is now due.

IV. SYMBOLIC INTERACTIONIST PERSPECTIVES ON DEVIANCE

A. Differential Association Theory and Differential Reinforcement Theory

1. **Differential association theory** states that individuals have a greater tendency to deviate from societal norms when they frequently associate with persons who are more favorable toward deviance than conformity.

B. Control Theory: Social Bonding

1. **Social bond theory** holds that the probability of deviant behavior increases when a person's ties to society are weakened or broken.

C. Labeling Theory

1. **Labeling theory** states that deviants are those people who have been successfully labeled as such by others.
2. **Primary deviance** is the initial act of rule breaking.
3. **Secondary deviance** occurs when a person who has been labeled a deviant accepts that new identity and continues the deviant behavior.
4. **Tertiary deviance** occurs when a person who has been labeled a deviant seeks to normalize the behavior by labeling it as nondeviant.

V. POSTMODERNIST PERSPECTIVES ON DEVIANCE

A. Postmodernists perspective on deviance examines the intertwining nature of knowledge, power, and technology on social control and discipline.

B. Michael Foucault's view on deviance and social control has influenced other social analysts, including Shoshana Zuboff, who views the computer as a modern *Panoptican* that gives workplace supervisors virtually unlimited capabilities for surveillance over subordinates.

VI. CRIME CLASSIFICATIONS AND STATISTICS

A. How the Law Classifies Crime

1. Crimes are divided into felonies and misdemeanors based on the seriousness of the crime.

B. Other Crime Categories

1. The **Uniform Crime Report** is the major source of information on crimes and is compiled by the Federal Bureau of Investigation.
   a. **Violent crime** consists of actions- murder, forcible rape, robbery, and aggravated assault- involving force or the threat of force against others.
   b. **Property crime**- includes burglary, motor vehicle theft, larceny theft, and arson.
   c. **Public order crime**- involves an illegal action voluntarily engaged in by participants, such as prostitution or illegal gambling.
   d. **Occupational** and **corporate crime** involves illegal activities committed by people in the course of their employment or financial affairs.
   e. **Organized crime** is a business operation that supplies illegal goods and services for profit.
   f. **Political crime** refers to illegal or unethical acts involving the usurpation of power by government officials, or illegal/unethical acts perpetrated against the government by outsiders seeking to make a political statement, undermine the government, or overthrow it.

C. Crime Statistics

1. Official crime statistics, such as those found in the *Uniform Crime Report*, provide important information on crime; however, the data reflect only those crimes that have been reported to the police.

2. The National Crime Victimization Survey and anonymous self-reports of criminal behavior have made researchers aware that the incidence of some crimes, such as theft, is substantially higher than reported in the UCR.

3. Crime statistics do not reflect many crimes committed by persons of upper socioeconomic status in the course of business because they are handled by administrative or quasi-judicial bodies.

D. Terrorism and Crime
   1. **Terrorism** is the calculated, unlawful use of physical force or threats of violence against persons or property in order to intimidate or coerce a government, organization, or individual for the purpose of gaining some political, religious, economic, or social objective.
   2. The United States and other nations are confronted with world terrorism and crime.
   3. The nebulous nature of the "enemy" and other limitations has resulted in a "war on terrorism."

E. Street Crimes and Criminals
   1. Gender and Crime
      a. The three most common arrest categories for both men and women are driving under the influence of alcohol or drugs (DUI), larceny, and minor or criminal mischief types of offenses.
      b. Liquor law violations (such as underage drinking), simple assault, and disorderly conduct are middle range offenses for both men and women, and the rate of arrests for murder, arson, and embezzlement are relatively low for both men and women.
      c. There is a proportionately greater involvement of men in major property crimes and violent crime.
   2. Age and Crime
      a. Arrest rates for index crimes are highest for people between the ages of 13 and 25, with the peak being between ages 16 and 17.
      b. In 2004, persons under age 25 accounted for more than 44 percent of all arrests for violent crime and almost 60 percent of all arrests for property crime.
      c. Individuals under age 18 accounted for over 23 percent of all arrests for robbery and 27 percent of all arrests for larceny theft.
      d. Rates of arrest remain higher for males than females at every age and for nearly all offenses.
   3. Social Class and Crime
      a. Individuals from all social classes commit crimes; they simply commit different kinds of crime.
      b. Persons from lower socioeconomic backgrounds are more likely to be arrested for violent and property crimes; only a very small proportion of individuals who commit white-collar or elite crimes will ever be arrested or convicted.

4. Race and Crime
   a. In 2004, whites (including Latinos/as) accounted for about 70 percent of all arrests; arrest rates for whites were higher in non-violent property crimes such as fraud and larceny-theft, but were lower than the rates for African Americans in violent crimes such as robbery and murder.
   b. In 2004, African Americans made up about 12 percent of the U.S. population but accounted for almost 27 percent of all arrests.
   c. Latinos/as made up about 13 percent of the U.S. population and accounted for about 13 percent of all arrests; over two-thirds of their offences were for non-index crimes such as alcohol and drugs-related offences and disorderly conduct.
   d. In 2004, about 1 percent of all arrests were Asian-Americans or Pacific Islanders.
   e. About 1 percent of all arrests were Native Americans, designated in the UCR as "American Indian" or "Alaskan Native."
   f. Arrest records tend to produce over generalizations about who commits crime because arrest statistics are not an accurate reflection of the crimes actually committed in our society.
   g. Arrests should not be equated with guilt: being arrested does not mean that a person is guilty of the crime.
5. Crime Victims
   a. Men are more likely to be victimized by crime although women tend to be more fearful of crime, particularly those directed toward them, such as forcible rape.
   b. The elderly also tend to be more fearful of crime, but are the least likely to be victimized. Young men of color between the ages of 12 and 24 have the highest criminal victimization rates.
   c. The burden of robbery victimization falls more heavily on males than females, African Americans more than whites, and young people more than middle-aged and older persons.

VII. CRIMINAL JUSTICE SYSTEM

A. The criminal justice system includes the police, the courts, and prisons. This system is a collection of bureaucracies that has considerable discretion -- the use of personal judgment regarding whether to take action on a situation and, if so, what kind of action to take.

B. The **police** are responsible for crime control and maintenance of order.

C. The **courts** determine the guilt or innocence of those accused of committing a crime.

D. **Punishment** is any action designed to deprive a person of things of value (including liberty) because of something the person is thought to have done.

1. Punishment is seen as serving four functions: retribution, general deterrence, incapacitation, and rehabilitation.

2. Disparate treatment of the poor, people of color, and women is evident in the prison system.

E. For many years, capital punishment, or the death penalty, has been used in the United States; about 4,000 executions have occurred in the U.S. since 1930, and scholars have documented race and class biases in the imposition of the death penalty in this country.

VIII. DEVIANCE AND CRIME IN THE UNITED STATES IN THE FUTURE

A. Although many people in the United States agree that crime is one of the most important problems facing this country, they are divided over what to do about it.

B. The best approach for reducing delinquency and crime ultimately is prevention: to work with young people before they become juvenile offenders so as to help them establish family relationships, build self-esteem, choose a career, and get an education which will help them pursue that career, as well as to promote social justice regardless of race, class, gender, or age.

IX. THE GLOBAL CRIMINAL ECONOMY

A. Global Crime--the networking of powerful criminal organizations and their associates in shared activities around the world-have expanded rapidly in the era of global communications and rapid transportation networks.

B. Global networking and forming strategic alliances allow participants to escape police control and live beyond the laws of any one nation.

C. Reducing global crime will require a global response, including the cooperation of law enforcement agencies around the world at the appropriate time.

ANALYZING AND UNDERSTANDING THE BOXES

After reading the chapter and studying the outline, re-read the boxes and write down key points and possible questions for class discussion.

Sociology and Everyday Life: How Much Do You Know About Peer Cliques, Youth Gangs, and Deviance?

Key Points:

Discussion Questions:

1.

2.

3.

You Can Make a Difference: Combating Delinquency and Crime at an Early Age

Key Points:

Discussion Questions:

1.

2.

3.

Sociology in Global Perspective: The Global Reach of Russian Organized Crime

Key Points:

Discussion Questions:

1.

2.

3.

PRACTICE TESTS

MULTIPLE CHOICE QUESTIONS

Select the response that best answers the question or completes the statement:

1. The text defines deviance as any: (p. 160)
   a. aberrant behavior.
   b. behavior, belief, or condition that violates cultural norms.
   c. serious violation of consistent moral codes.
   d. perverted act.

2. All of the following statements regarding deviance are TRUE, except: (p. 161)
   a. behavioral deviance is based on a person's intentional or inadvertent actions.
   b. people may be regarded as deviant if they express a radical or unusual belief system.
   c. people may be regarded as deviant because of specific characteristics or conditions that they have had since birth or have acquired.
   d. definitions of deviance are similar from place to place, from time to time, and from group to group.

3. A ________ is a behavior that violates criminal law and is punishable with negative sanctions. (p. 162)
   a. stigma
   b. deviant act
   c. crime
   d. cultural strain

4. The systematic practices developed by social groups to encourage conformity and to discourage deviance are known as: (p. 162)
   a. laws.
   b. folkways.
   c. mores.
   d. social control.

5. __________ refers to a violation of law or the commission of a status offense by young people. (p. 162)
   a. Juvenile delinquency
   b. Truancy
   c. Youthful misconduct
   d. Crime

6. The police, the courts, and the prisons are examples of: (p. 163)
   a. internal social control.
   b. moral policing.
   c. external social control.
   d. mores.

7. Which of the following is more likely to take place through the socialization process? (p. 162)
   a. external social control
   b. mild social control
   c. civil social control
   d. internal social control

8. The systematic study of crime and the criminal justice system, including the police, courts, and prisons is termed: (p. 163)
   a. sociology.
   b. criminology.
   c. ecology.
   d. the criminal justice system.

9. According to sociologist __________, deviance is a natural and inevitable part of all societies. (p. 163)
   a. Robert Merton
   b. Emile Durkheim
   c. Karl Marx
   d. Walter Reckless

10. According to functionalists such as Emile Durkheim, deviance serves all of the following functions, *except*: (pp. 163-164)
   a. deviance helps us to identify social dynamite and social junk in a society.
   b. deviance clarifies rules.
   c. deviance promotes social change.
   d. deviance unites a group.

11. Suppose that a bully comes into town and the residents band together in order to oppose the threat. This illustrates which function of deviance? (pp. 163-164)
   a. Deviance clarifies rules.
   b. Deviance unites a group.
   c. Deviance promotes social change.
   d. Deviance prevents social chaos.

12. According to the text, acts of civil disobedience exemplify which function of deviance? (pp. 163-164)
   a. Deviance clarifies rules.
   b. Deviance unites a group.
   c. Deviance promotes social change.
   d. Deviance prevents social chaos.

13. According to _____ theory, people are frustrated when they are exposed to cultural goals that they are unable to obtain because they do not have access to culturally approved means of achieving those goals. (p. 164)
   a. containment
   b. status inaccessibility
   c. strain
   d. conflict

14. All of the following are included in Robert Merton's modes of adaptation to cultural goals and approved ways of achieving them, *except*: (pp. 164-165)
    a. retribution.
    b. ritualism.
    c. retreatism.
    d. rebellion.

15. Based on Robert Merton's typology, a government service employee who adheres to the established rules so completely that she or he often loses sight of the agency's purpose is engaged in: (pp. 164-165)
    a. retribution.
    b. ritualism.
    c. retreatism.
    d. rebellion.

16. A young woman graduates from high school with honors and attends a prestigious university, where she completes her degree; she gets a good job; she marries and starts planning for the future. This woman's behavior illustrates: (p.164)
    a. conformity.
    b. innovation.
    c. rebellion.
    d. rejection.

17. According to _____ theory, a teenager living in a poverty-ridden area of a central city is unlikely to become wealthy through a Harvard education, but some of his or her desires may be met through behaviors such as theft, drug dealing, and robbery. (p. 165)
    a. deviance management
    b. control
    c. illegitimate opportunity structures
    d. critical

18. _____ theories suggest that the probability of delinquency increases when a person's social bonds are weak and when peers promote antisocial values and violent behavior. (p. 170)
    a. Deviance management
    b. Control
    c. Illegitimate opportunity structures
    d. Critical

19. According to Edwin Lemert's typology, _____ deviance is exemplified by a person under the legal drinking age who orders an alcoholic beverage at a local bar but is not "caught" and labeled a deviant. (p. 171)
    a. primary
    b. secondary
    c. residual
    d. adolescent

20. The _____ approach argues that criminal law protects the interests of the affluent and powerful. (p. 167)
    a. functionalist
    b. liberal feminist
    c. symbolic interactionist
    d. conflict

21. _____ feminism explains women's deviance and crime as a rational response to gender discrimination experienced in work, marriage, and interpersonal relationships. (p. 168)
    a. Radical
    b. Communist
    c. Liberal
    d. Marxist (socialist)

22. Socialist feminists argue that women's deviance and crime occurs because: (p. 168)
    a. women are exploited by other women.
    b. women are exploited by capitalism and patriarchy.
    c. women experience gender discrimination in work, marriage, and interpersonal relationships.
    d. women are consumers and tend to purchase more than they can afford.

23. A _____ is a serious crime for which punishment typically ranges from more than a year's imprisonment to death. (p. 172)
    a. misdemeanor
    b. property crime
    c. felony
    d. morals crime

24. Violent crime and certain property crimes are referred to as _____ crime by the 2004 UCR Report. (p. 173)
    a. misdemeanor
    b. organized
    c. political
    d. index

25. All of the following are index crimes, *except*: (pp. 173-174)
    a. traffic violations.
    b. larceny in excess of $50.
    c. armed robbery.
    d. murder.

26. Occupational crime is also known as _____ crime. (p. 175)
    a. white-collar
    b. street
    c. organized
    d. conventional

27. Drug trafficking, prostitution, loan-sharking, and money-laundering are examples of _____ crime. (p. 176)
    a. white-collar
    b. street
    c. organized
    d. conventional

28. All of the following are examples of political crime, *except*: (pp. 176-177)
    a. unethical or illegal use of government authority for the purpose of material gain.
    b. money-laundering.
    c. engaging in graft through bribery, kickbacks, or "insider" deals.
    d. dubious use of public funds and public property.

29. According to the text, rates of arrest: (p. 177)
    a. are about the same for males and females at every age group and for most offenses.
    b. are slightly higher for females than males in the younger age levels and for violent crimes.
    c. are about the same for males and females for prostitution due to more stringent enforcement of criminal laws pertaining to male customers.
    d. remain higher for males than females at every age for violent crimes.

30. All of the following are functions of punishment, *except*: (pp. 184-185)
    a. deterrence.
    b. retribution.
    c. incapacitation.
    d. elimination of social dynamite and social junk.

TRUE-FALSE QUESTIONS

T F 1. People may be regarded as deviant if they express a radical or unusual belief system. (p. 160)

T F 2. According to sociologists, deviance is relative. (p. 161)

T F 3. Juvenile delinquency includes status offenses. (p. 162)

T F 4. Primary/secondary deviance and labeling theory are functionalist perspectives on deviance and crime. (pp. 170-171)

T F 5. According to differential association theory, people learn the necessary techniques and motivation for deviant behavior from people with whom they associate. (p. 169)

T F 6. The concept of secondary deviance is important because it suggests that when people accept a negative label or stigma that has been applied to them, the label may contribute to the type of behavior it initially was meant to control. (p. 171)

T F 7. Karl Marx wrote extensively about deviance and crime. (p. 167)

T F 8. According to the emancipation theory of female crime, women's crime rates are expect to decrease significantly. (pp. 167-168)

T F 9. Recently, functionalist theorists have examined the relationship between class, race, gender, and crime. (p. 168)

T F 10. Terrorism is the calculated, unlawful use of physical force or threats of violence against persons or property in order to intimidate or coerce a government, organization, or individual for the purpose of gaining some political, religious, economic, or social objective. (p. 178)

T F 11. Postmodernist perspectives on deviance examine the intertwining nature of power, knowledge, and social control. (p. 172)

T F 12. Organized crime groups engage only in illegal enterprises. (p. 176)

T F 13. Official crime statistics provide very accurate data on the number of crimes committed each year in the United States. (p. 177)

T F 14. Among males who are twelve years old, an estimated 89 percent will be victims of a violent crime at least once during their lifetime. (p. 181)

T F 15. Ex-slave states are less likely to execute criminals than are other states. (p. 186)

SOCIOLOGY IN OUR TIMES: DIVERSITY ISSUES

1. Gang membership provides some men and women in low-income, central-city areas with illegitimate means to achieve conventional goals of status and wealth. What are other ways that society might help these young people meet their needs other than through criminal activity?

2. As a category, do white women and people of color have the same opportunities to commit high-level occupational and corporate crimes as white men? Does your answer tell us anything about people's access (or lack of access) to high-paying, prestigious careers and professions?

3. If you were going to study gender and crime, what topics might you explore?

4. What are some of illegal income producing activities of the global criminal economy? How do these illegal activities make it possible for some to move into the "legitimate" formal economy?

5. How are class and gender related to crime statistics?

6. Who are the most frequent victims of crime?

7. How is discretion used in the criminal justice system?

INTERNET EXERCISES

Access the Internet and use a search engine to complete the following:

1. Give an example, aside from the Mafia, of organized crime that currently exists.

2. Read a variety of articles on the death penalty and then decide whether or not it is a deterrent of crime. If so, which crimes does it deter people from engaging in; and if not, what would deter people?

INFOTRAC COLLEGE EDITION ONLINE READINGS AND EXERCISES

Log onto the Infotrac website. In the **keyword** search field enter "National Center for Missing and Exploited Children Remembers AMBER Alert Namesake" and hit search. Then select the article entitled: "National Center for Missing and Exploited Children Remembers AMBER Alert Namesake." Read the article and answer the following questions:

a. What does AMBER stand for?

b. What information do the broadcasts include?

Now return to the **keyword** search field and enter "Mafia blamed for soaring Italian food prices." Click on the title "Mafia blamed for soaring Italian food prices". Read the article and answer the following questions:

a. How has the mafia affected food prices for consumers?

b. What campaign have the farmers launched in order to help with the problem?

STUDENT CLASS PROJECTS AND ACTIVITIES

1. The organized crime syndicate, the Mafia, has a long and interesting history. You are to research the origin, the purpose, and the growth of the Mafia in Italy and Sicily, noting some similarities and differences of each. In addition to tracing the history of both organizations, explain the growth of the Cupoda in Sicily. Explain some of the original neighborhood programs which the Mafia sponsored. Explain why you think it is difficult to eradicate this crime syndicate.

2. Research the topic of crime rates in a cross-cultural perspective, with specific emphasis on comparing the rates in the U.S. with those in other industrialized countries. Summarize your findings, provide some explanations for your findings, and provide a bibliography of your references. Submit your papers at the appropriate time, following the instructions you have been given in the writing of your paper.

3. As indicated in your textbook, global crime--the networking of powerful criminal organizations and their associates in shared activities around the world--is a relatively new phenomenon, although theses organizations have existed for many years in their country of origin. You are to: 1) select five global criminal organizations; 2) explain the origin, purpose, and growth of the organization; 3) describe some of their illegal activity; 4) examine and explain some key factors in the success of expansion of their criminal activity; 5) provide some possible solutions for reducing global crime; and 6) critique your solutions. What would it take for your solutions to work?

4. Research the identification and expansion of world terrorism. What nations of the world today support and/or sponsor world terrorism? Why? In your research, identify various diverse "cells" of terrorist. Who are their leaders? What theoretical approaches best explain the existence of terrorists? What are some possible solutions for eradicating world terrorism and crime? Provide bibliographic references in your writing up this project. Follow any additional instructions provided by your professor and submit your paper at the appropriate time

ANSWERS TO PRACTICE TESTS FOR CHAPTER 6

Answers to Multiple Choice Questions

1.	b	The text defines deviance as any behavior, belief, or condition that violates cultural norms. (p. 160)
2.	d	Deviance varies according to the culture (p. 161)
3.	c	formal definition. (p. 162)

4. d Social control refers to systematic practices developed by social groups to encourage conformity and to discourage deviance. (p. 162)
5. a Young people equates to juveniles. (p.162)
6. c These three are the primary types of external social control. (p.163)
7. d This occurs through socialization. (p. 163)
8. b The academic discipline. (p. 163)
9. b This is the common functionalist perspective. (p. 163)
10. a According to functionalists such as Emile Durkheim, deviance serves all of the following functions, *except*: deviance helps us to identify social dynamite and social junk in a society. (p. 163-164)
11. b This is relatively common in societies, especially in war. (p. 164)
12. c Social change can occur through deviance. (pp. 163-164)
13. c According to strain theory, people are frustrated when they are exposed to cultural goals that they are unable to obtain because they do not have access to culturally approved means of achieving those goals. (p. 164)
14. a All of the following are included in Robert Merton's modes of adaptation to cultural goals and approved ways of achieving them, *except* retribution. (pp. 164-165)
15. b Based on Robert Merton's typology, a government service employee who adheres to the established rules so completely that she or he often loses sight of the agency's purpose is engaged in ritualism. (pp. 164-165)
16. a This is the desired end. (p. 164)
17. c According to illegitimate opportunity structures theory, a teenager living in a poverty-ridden area of a central city is unlikely to become wealthy through a Harvard education, but some of his or her desires may be met through behaviors such as theft, drug dealing, and robbery. (p. 165)
18. b Control theories suggest that the probability of delinquency increases when a person's social bonds are weak and when peers promote antisocial values and violent behavior. (p. 170)
19. a According to Edwin Lemert's typology, primary deviance is exemplified by a person under the legal drinking age who orders an alcoholic beverage at a local bar but is not "caught" and labeled a deviant. (p. 171)
20. d The conflict approach argues that criminal law protects the interests of the affluent and powerful. (p. 167)
21. c Liberal feminism explains women's deviance and crime as a rational response to gender discrimination experienced in work, marriage, and interpersonal relationships. (p. 168)
22. b Marxist (socialist) feminists argue that women's deviance and crime occurs because women are exploited by capitalism and patriarchy. (p. 168)
23. c A felony is a serious crime for which punishment typically ranges from more than a year's imprisonment to death. (p. 172)
24. d Violent crimes and certain property crimes are referred to as index crimes by the 2004 UCR Report. (p. 173)
25. a All of the following are index crimes, *except* traffic violations. (pp. 173-174)
26. a Occupational crime is also known as white-collar crime. (p. 175)

27. c Drug trafficking, prostitution, loan-sharking, and money-laundering are examples of organized crime. (p. 176)
28. b All of the following are examples of political crime, *except* money-laundering. (pp. 176-177)
29. d According to the text, arrest rates remain higher for males than females at every age for violent crimes. (p. 177)
30. d All of the following are functions of punishment, *except*: elimination of social dynamite and social junk. (pp. 184-185)

Answers to True-False Questions

1. True (p. 160)

2. True (p. 161)

3. True (p. 162)

4. False Functionalist perspectives include strain theory, opportunity theory, and social control/social bonding. Primary/secondary deviance and labeling theory are symbolic interactionist perspectives. (pp. 170-171)

5. True (p. 169)

6. True (p. 171)

7. False Although Marx wrote very little about deviance and crime, many of his ideas are found in a critical approach that has emerged from earlier Marxist and radical perspectives on criminology. (p. 167)

8. False The emancipation theory of female crimes declares that women's crime rates would increase significantly as a result of the women's liberation movement. (pp. 167-168)

9. False Recently, critical conflict theorists have examined the relationship between class, race, gender, and crime. (p. 168)

10. True (p. 178)

11. True (p. 172)

12. False Organized crime groups have infiltrated the world of legitimate business, such as banking, real estate, garbage collection, and garment manufacturing. (p. 176)

13. False Official crime statistics reflect only those crimes that have been reported to the police; victimization surveys indicate that the incidence of some crimes is substantially higher than reported in official crime reports. (p. 177)

14. True (p. 181)

15. False Ex-slave states are more likely to execute criminals than are other states. (p. 186)

CHAPTER 7

CLASS AND STRATIFICATION IN THE UNITED STATES

BRIEF CHAPTER OUTLINE

What Is Social Stratification?
- **Global Systems of Stratification**
  - **Slavery**
  - The Caste System
  - The Class System
- **Classical Perspectives on Social Class**
  - Karl Marx: Relationship to the Means of Production
  - Max Weber: Wealth, Prestige, and Power
- **Contemporary Sociological Models of the U.S. Class Structure**
  - The Weberian Model of the U.S. Class Structure
  - The Marxian Model of the U.S. Class Structure
- **Inequality in the United States**
  - Distribution of Income and Wealth
  - Consequences of Inequality
- **Poverty in the United States**
  - Who Are the Poor?
  - Economic and Structural Sources of Poverty
  - Solving the Poverty Problem
- **Sociological Explanations of Social Inequality in the United States**
  - Functionalist Perspectives
  - Conflict Perspectives
  - Symbolic Interactionist Perspectives
- **U.S. Stratification in the Future**

CHAPTER SUMMARY

Social stratification is the hierarchical arrangement of large social groups based on their control over basic resources. A key characteristic of systems of stratification is the extent to which the structure is flexible. **Social mobility** is the movement of individuals or groups from one level in a stratification level to another. **Slavery**, a form of stratification in which people are owned by others, is a closed system. In a **caste system**, people's status is determined at birth based on their parents' position in society. The **class system**, which exists in the United States, is a type of stratification based on ownership of resources and on the type of work that people do. Karl Marx and Max Weber viewed class as a key determinant of social inequality and social change. According to Marx, capitalistic societies are comprised of two classes—the capitalists, who own the means of production, and the workers, who sell their labor to the owners. By contrast, Weber developed a multidimensional concept that focuses on the interplay of **wealth, prestige**, and **power**. Sociologists have developed several models of the class structure: one is the broadly based Weberian approach; the second is based on a Marxian approach. Throughout human history, people have argued about the distribution of scarce resources in society. Money, in the form of both income and wealth is unevenly distributed in the U. S. Among the most prosperous nations in today's world, the United States has the highest degree of inequality of income distribution. **Income** –the economic gain derived from wages, salaries, income transfers, or ownership of property—and **wealth**—the value of all of a person's or family's economic aspects—is unevenly distributed in the United States. The stratification of society into different social groups results in wide discrepancies in income, wealth, and access to available goods and services (including physical health, mental health, nutrition, housing, education, and safety). The U.S. Social Security Administration has established an **official poverty line,** which is based on what is considered to be the minimum amount of money required for living at a subsistence level. Sociologists distinguish between **absolute poverty**, which exists when people do not have the means to secure the basic necessities of life, and **relative poverty**, which exists when people may be able to afford basic necessities but are still unable to maintain an average standard of living. Poverty is highly concentrated according to age, gender and race/ethnicity. There are both economic and structural sources of poverty. Low wages are a key problem, as are unemployment and underemployment. The United States has attempted to solve the poverty problem, with the most enduring one referred to as social welfare. Functionalists' perspectives on the U.S. class structure view classes as broad groupings of people who share similar levels of privilege based on their roles in the occupational structure. According the Davis-Moore thesis, positions that are most important within a society, requiring the most talent and training, must be highly rewarded. Conflict perspectives are based on the assumption that social stratification is created and maintained by one group in order to enhance and protect its own economic interests. Symbolic interactionists focus on the micro-level analysis; such as examining the social and psychological factors that influence the contributions of the wealthy to charitable and arts organizations. As the gap between rich and poor, employed and unemployed widens, social inequality will increase in the future if society does not address this problem. Given that the well-being of all people is linked, some analysts are urging a joint effort to regain the American Dream by attacking poverty.

LEARNING OBJECTIVES

After reading Chapter 7, you should be able to:

1. Define social stratification and describe the major sources of stratification found in societies.
2. Define life chances and explain its effects upon people.
3. Explain social mobility and distinguish between intergenerational and intragenerational mobility.
4. Describe global systems of stratification, including the key characteristics of the three major systems of stratification.
5. Describe Karl Marx's perspective on class position and class relationships.
6. Outline Max Weber's multidimensional approach to social stratification and explain how people are ranked on all three dimensions.
7. Outline the functionalist model of the U.S. class structure and briefly describe the key characteristics of each class.

8. Outline the conflict model of the U.S. class structure and briefly describe the key characteristics of each class.
9. Identify and compare the terms income and wealth.
10. Describe some of the benefits of wealth.
11. Compare and contrast the distribution of income and wealth in the United States.
12. Discuss the distribution of income and wealth in the United States and describe how this distribution affects life chances.
13. Define the term, official poverty line. Provide an example.
14. Distinguish between absolute and relative poverty and describe the characteristics and lifestyle of those who live in poverty in the United States.
15. Describe the feminization of poverty and explain why two out of three impoverished adults in the United States are women.
16. Discuss some economic and structural sources of poverty.
17. Provide some of the attempts of the United States to solve the poverty problem.
18. Distinguish between functionalist and conflict explanations of social inequality.
19. Explain the symbolic interactionist perspectives of social inequality in the United States.

KEY TERMS (defined at page number shown and in glossary)

absolute poverty 216
alienation 199
capitalist class (or bourgeoisie) 199
caste system 198
class conflict 199
class system 198
Davis-Moore thesis 220
feminization of poverty 218
income 208
intergenerational mobility 194
intragenerational mobility 194
job deskilling 219
life chances 194
meritocracy 220
official poverty line 216
pink collar occupations 204
power 201
prestige 201
relative poverty 216
slavery 195
social mobility 194
social stratification 194
socioeconomic status (SES) 201
underclass 205
wealth 201

CHAPTER OUTLINE

I. WHAT IS SOCIAL STRATIFICATION?
   A. **Social stratification** is the hierarchical arrangement of large social groups based on their control over basic resources.
   B. Max Weber's term **life chances** describes the extent to which persons within a particular layer of stratification have access to important scarce resources.

II. GLOBAL SYSTEMS OF STRATIFICATION

A. Systems of stratification may be open or closed based on the availability of **social mobility**—the movement of individuals or groups from one level in a stratification system to another.

1. **Intergenerational mobility** is the social movement experienced by family members from one generation to the next.
2. **Intragenerational mobility** is the social movement of individuals within their own lifetime

B. **Slavery**, a closed system, is an extreme form of stratification in which some people are owned by others.

1. Some analysts suggest that throughout recorded history, only five societies have been slave societies.
2. Sociologist Collins suggests that the legacy of slavery is embedded in current patterns of prejudice and discrimination against African Americans.
3. Engerman belives that slavery will not be totally abolished until debt bondage, child labor, contract labor and coerced work cease to exist throughout the world.

C. A **caste system** is a system of social inequality in which people's status is permanently determined at birth based on their parents' ascribed characteristics.

D. The **class system** is a type of stratification based on the ownership and control of resources and on the type of work people do.

III. CLASSICAL PERSPECTIVES ON SOCIAL CLASS

A. Karl Marx: Relationships to the Means of Production

1. According to Marx, class position in capitalistic societies is determined by people's work situation, or relationship to the means of production.
   a. The **bourgeoisie** or **capitalist class** consists of those who privately own the means of production; the **proletariat**, or **working class**, must sell their labor power to the owners in order to earn enough money to survive.
   b. Class relationships involve inequality and exploitation; workers are exploited as capitalists expropriate a surplus value from their labor; continual exploitation results in workers' **alienation**, a feeling of powerlessness and estrangement from other people and from oneself.
2. The **capitalist class** maintains its position by control of the society's superstructure—comprised of the government, schools, and other social institutions, which produce and disseminate ideas perpetuating the existing system; this exploitation of workers ultimately results in **class conflict**—the struggle between the capitalist class and working class.

B. Max Weber: Wealth, Prestige, and Power

1. Weber's multidimensional approach to stratification examines the interplay among wealth, prestige, and power as being necessary in determining a person's class position.

a. Weber placed people who have a similar level of **wealth**—the value of all of a person or family's economic assets, including income, personal property, and *income* producing property—and income in the same class.
b. **Prestige** is the respect or regard with which a person or status position is regarded by others, and those who share similar levels of social prestige belong to the same status group regardless of their level of wealth.
c. **Power**—the ability of people or groups to carry out their own goals despite opposition from others—gives some people the ability to shape society in accordance with their own interests and to direct the actions of others.

2. Wealth, prestige, and power are separate continuums on which people can be ranked from high to low; individuals may be high on one dimension while being low on another.
3. **Socioeconomic status (SES)** -- a combined measure that attempts to classify individuals, families, or households in terms of indicators such as income, occupation, and education—is used to determine class location.

IV. CONTEMPORARY SOCIOLOGICAL MODELS OF THE U.S. CLASS STRUCTURE

A. The Weberian Model of the Class Structure

1. The Upper (or Capitalist) Class is the wealthiest and most powerful class, comprised of people who own substantial income producing assets. About 1 percent of the population is included in this class.
2. The Upper-Middle Class is based on a combination of three factors: university degrees, authority and independence on the job, and high income. Examples of occupations for this class are highly educated professionals such as physicians, stockbrokers, or corporate managers. About 14 percent of the U. S. population is in this category.
3. The Middle Class has been traditionally characterized by a minimum of a high school diploma. Today, an entry-level requirement for employment in many middle-class occupations requires a two-year or four-year college degree. About 30 percent of the U. S. population is in this class.
4. The Working Class is comprised of semiskilled machine operatives, clerks and salespeople in routine, mechanized jobs, and workers in **pink collar occupations**—relatively low paying, non-manual semiskilled positions primarily held by women. An estimated 30 percent of the U. S. population is in this class.
5. The Working Poor account for about 20 percent of the U. S. population and live from just above to just below the poverty line; they hold unskilled jobs, seasonal migrant employment in agriculture, lower paid factory jobs, and service jobs (e.g., such as counter help at restaurants).

6. The **Underclass** includes people who are poor, seldom employed, and are caught in long term deprivation that results from low levels of education and income and high rates of unemployment. About 3 to 5 percent of the U. S. population is in this category.

B. The Marxian Model of the U.S. Class Structure

1. These criteria can be used to determine the class placement for all workers in a capitalist society.
   a. Erik Olin Wright outlined four criteria for placement in the class structure: (a) ownership of the means of production; (b) purchase of the labor of others (employing others); (c) control of the labor of others (supervising others on the job); and (d) sale of one's own labor (being employed by someone else).
2. The Capitalist Class is composed of those who have inherited fortunes, own major corporations, or are top corporate executives who own extensive amounts of stock or control company investments.
3. The Managerial Class includes upper level managers—supervisors and professionals who typically do not participate in company wide decisions—and lower level managers who may be given some control over employment practices, such as the hiring and firing of some workers.
4. The Small-Business Class consists of small business owners, craftspeople, and some doctors and lawyers who may hire a small number of employees but largely do their own work.
5. The Working Class is made up of blue collar workers, including skilled workers (e.g., electricians, plumbers, and carpenters), unskilled blue collar workers (e.g., laundry and restaurant workers), and white collar workers who do not own the means of production, do not control the work of others, and are relatively powerless in the workplace. The working class constitutes about half of all employees in the U. S.

V. INEQUALITY IN THE UNITED STATES

A. Physical and Mental Health and Nutrition

1. As people's economic status increases so does their health status; the poor have shorter life expectancies and are at greater risk for chronic illnesses and infectious diseases. About 41 million people in the United States are without health insurance coverage.

B. Housing

1. Homelessness is a major problem in the U.S.
2. Lack of affordable housing and substandard housing are also central problems brought about by economic inequality.

C. Education

1. Education and life chance are directly linked; while functionalists view education as an "elevator" for social mobility, conflict theorists stress that schools are agencies for reproducing the capitalist class system and perpetuating inequality in society.

2. Poverty affects the ability of many young people to finish high school, much less enter college.

D. Crime and Lack of Safety

1. Both crime and lack of safety on the streets and at home are consequences of inequality.
2. Poverty and violence are linked; street violence is often not random at all—but a response to profound social inequalities in the inner city.

VI. POVERTY IN THE UNITED STATES

A. Athough some people living in poverty are unemployed, many hardworking people with full time jobs also live in poverty.

B. The official poverty line is based on what is considered to be the minimum amount of money required for living at a subsistence level.

C. Sociologists distinguish between **absolute poverty**—when people do not have the means to secure the most basic necessities of life—and **relative poverty**—when people may be able to afford basic necessities but still are unable to maintain an average standard of living.

D. Are the Poor?

1. Age: Children are more likely to be poor than older persons; older women are twice as likely to be poor as older men; older African Americans and Latinos/as are much more likely to live below the poverty line than are non Latino/a whites.
2. Gender: About two thirds of all adults living in poverty are women; this problem is described as the **feminization of poverty**—the trend in which women are disproportionately represented among individuals living in poverty.
3. Race and Ethnicity: white Americans (non-Latinos/as) account for approximately two thirds of those below the official poverty line; however, a disproportionate percentage of the poverty population is made up of African Americans, Latinos/as, and Native Americans.

E. Economic and Structural Sources of Poverty

1. An economic source of poverty is the low wages paid for many jobs half of all families living in poverty are headed by someone who is employed, and one third of those family heads work full time.
2. Poverty also is exacerbated by structural problems such as (a) deindustrialization—millions of U.S. workers have lost jobs as corporations have disinvested here and opened facilities in other countries where "cheap labor" exists—and (b) **job deskilling**—a reduction in the proficiency needed to perform a specific job that leads to a corresponding reduction in the wages paid for that job.

F. Solving the Poverty Problem

1. The United States has attempted to solve the poverty problem withsocial welfare programs; however, the primary beneficiaries are not poor.
2. Recent major suggestions for solving the poverty problem are: (a) changing past welfare programs, in the name of "welfare reform and establish state-level workfare programs and mandatory time limits on

welfare benefits; and (b) providing better training and education in order that the poor might be able to "work their way out of poverty."

VII. SOCIOLOGICAL EXPLANATIONS OF SOCIAL INEQUALITY IN THE UNITED STATES

A. Functionalist Perspectives

1. According to the Davis-Moore thesis:
    a. All societies have important tasks that must be accomplished and certain positions that must be filled.
    b. Some positions are more important for the survival of society than others.
    c. The most important positions must be filled by the most qualified people.
    d. The positions that are the most important for society and require scarce talent, extensive training, or both, must be the most highly rewarded.
    e. The most highly rewarded positions should be those that are functionally unique (no other position can perform the same function), and those positions upon which others rely for expertise, direction, or financing.
2. This thesis assumes that social stratification results in **meritocracy**—a hierarchy in which all positions are rewarded based on people's ability and credentials.

B. Conflict Perspectives

1. From a conflict perspective, inequality does not serve as a source of motivation for people; powerful individuals and groups use ideology to maintain their favored positions at the expense of others.
2. Core values, laws, and informal social norms support inequality in the United States (e.g., legalized segregation and discrimination produce higher levels of economic inequality).credentials.

C. Symbolic Interactionist Perspectives

1. Symbolic interactionists focus on microlevel concerns such as the effects of wealth and poverty on people's lives.
2. Some studies focus on the social and psychological factors that influence the rich to contribute to charitable and arts organizations.
3. Other researchers have examined the social and psychological aspects of life in the middle class; few have examined rare insights into the social interactions between people from vastly different social classes.

VIII. U.S. SOCIAL STRATIFICATION IN THE FUTURE

A. According to some social scientists, wealth will become more concentrated at the top of the U.S. class structure; as the rich have grown richer, more people have found themselves among the ranks of the poor.

B. The gap between the earnings of workers and the income of managers and top executives in the U.S.A. has widened.

C. Structural sources of upward mobility are shrinking while the rate of downward mobility has increased; the persistence of economic inequality is related to profound global economic changes.

D. Some call for a united effort to regain the American Dream by attacking poverty.

ANALYZING AND UNDERSTANDING THE BOXES

After reading the chapter and studying the outline, re-read the boxes and write down key points and possible questions for class discussion.

Sociology in Everyday Life: How Much Do You Know About Wealth, Poverty, and the American Dream?
Key Points:

Discussion Questions:

1.

2.

3.

Sociology and Social Policy: Welfare Reform and It's Aftermath
Key Terms:

Discussion Questions:

1.

2.

3.

You Can Make a Difference: Feeding the Hungry
Key Terms:

Discussion Questions:

1.

2.

3.

PRACTICE TESTS

MULTIPLE CHOICE QUESTIONS

Select the response that best answers the question or completes the statement:

1. The "American Dream" is the belief that if people work hard and play by the rules, then they will have a chance to be successful. This idea is based on the assumption of __________. (p. 193)
   a. stratification
   b. inequality
   c. equality
   d. solidarity

2. Which of the following is a component of the "American Dream?" (p. 193)
   a. hard work will produce success
   b. each generat0ion will be more successful than the previous generation.
   c. everyone has an equal opportunity to succeed.
   d. all of the above.

3. According to sociologist Max Weber, the term __________ refers to the extent to which individuals have access to important societal resources such as food, clothing, shelter, education, and health care. (p. 194)
   a. meritocracy
   b. alienation
   c. intergenerational mobility
   d. life chances

4. Sociologists use the term _____ to refer to the hierarchical arrangements of large social groups based on their control over basic resources. (p. 194)
   a. social stratification
   b. social layering
   c. social distinction
   d. social accumulation

5. The extent to which individuals have access to important societal resources is known as: (p. 194)
   a. relative poverty.
   b. absolute poverty.
   c. social mobility.
   d. life chances.

6. A young woman's father is a carpenter; she graduates from college with a degree in accounting, becomes a CPA, and has a starting salary that represents more money than her father ever made in one year. This illustrates _____ mobility. (p. 194)
   a. intragenerational
   b. intergenerational
   c. horizontal
   d. subjective

7. All of the following are **true** statements about slavery, **except**: (pp. 195-197)
   a. Slavery is a closed system in which "slaves" are treated as property.
   b. Slaves were forcibly imported to the United States as a source of cheap labor.
   c. Slavery has ended throughout the world.
   d. Some people have been enslaved because of unpaid debts, criminal behavior, or war and conquest.

8. A _____ system is a system of social inequality in which people's status is permanently determined at birth based on their parents' ascribed characteristics. (p. 198)
   a. class
   b. slavery
   c. capitalist
   d. caste

9. A young woman who comes from an impoverished background works at two full-time jobs in order to save enough money to attend college. Ultimately, she earns a degree, attends law school, graduates with highest honors, and is hired by a firm at a starting salary of $75,000. This person has experienced _____ mobility. (p. 198)
   a. vertical, intragenerational
   b. vertical, intergenerational
   c. horizontal, class
   d. direct, vertical

10. According to _____'s theory of class relations: (p. 199)
   a. Weber; the bourgeoisie consists of those who own the means of production.
   b. Marx; the proletariat consists of those who own the means of production.
   c. Marx; class relationships involve inequality and exploitation.
   d. Durkheim; wealth, prestige, and power are all important in determining a person's class position

11. According to Karl Marx ___________ is a feeling of powerlessness and estrangement from other people and from oneself. (p. 199)
    a. alienation
    b. meritocracy
    c. class conflict
    d. classism

12. When workers overthrow the capitalists, according to Marx, they would eventually create a/an __________ society. (p. 200)
    a. class
    b. caste
    c. egalitarian
    d. stratified

13. According to the _____perspective, the explanation of social inequality in the U.S. is _______ (p. 221)
    a. symbolic interactionist; that the beliefs and actions of people reflect their class location in society.
    b. functionalist; that powerful groups use ideology to maintain their favored position at the expense of others.
    c. conflict ; that some degree of social inequality is necessary for society to flow smoothly.
    d. absolute poverty; in comparison with the social status of others.

14. All of the following statements regarding Marx's analyses of class are correct, **except**: (pp.199-200)
    a. Class relationships involve inequality and exploitation.
    b. The exploitation of workers by the capitalist class ultimately will lead to the destruction of capitalism.
    c. The capitalist class maintains its position by control of the society's superstructure.
    d. Wealth, prestige, and power are separate continuums on which people can be ranked from high to low.

15. According to Max Weber, _____ is the respect or regard with which a person or status position is regarded by others. (p. 201)
    a. admiration
    b. power
    c. prestige
    d. rank

16. Entrepreneurs, as identified by Weber are those who:
    a. do not have to work. (p. 201)
    b. work for wages.
    c. are a privileged commercial class.
    d. live off their investments.

17. In 2004, ___ people in the United States were without health insurance coverage. (p. 211)
    a. 25.2 million
    b. 45. 8 million
    c. 31.5 million
    d. 35 million

18. According to the Weberian model of the U.S. class structure, members of the _____ class have earned most of their money in their own lifetime as entrepreneurs, presidents of corporations, top level professionals, and so forth. (p. 203)
    a. upper upper
    b. lower upper
    c. upper middle
    d. middle

19. The text points out that a combination of three factors qualifies people for the upper middle class. Which of the following is not one of these factors?
    a. university degrees. (p. 203)
    b. authority and independence on the job.
    c. inherited wealth.
    d. high income.

20. Over the past fifty years, Asian Americans, Latinos/as, and African Americans have placed great emphasis on _____ as a means of attaining the American Dream. (p. 203)
    a. education
    b. affirmative action
    c. unemployment compensation
    d. vocational training

21. Women employed in pink collar occupations are mainly classified in the: (p. 204)
    a. working class.
    b. working poor.
    c. middle class.
    d. upper middle class.

22. In the Weberian Model of the U.S. Class Structure, medical technicians, nurses, lower-level managers, and semi-professionals make up the _______ class. (p. 203)
    a. upper
    b. middle
    c. working
    d. lower

23. The trend in which women disproportionately are represented among individuals living in poverty is referred to as: (p.201)
    a. absolute poverty.
    b. relative poverty.
    c. situational poverty.
    d. the feminization of poverty.

24. All of the following are components of wealth, **except**: (p. 201)
    a. income
    b. a position on the local school board
    c. bank accounts
    d. insurance policies

25. Employed single mothers often belong to _____ (p. 205)
    a. the upper middle class
    b. the middle class
    c. the working class
    d. the working poor

26. Erick Olin Wright suggests that there are four social classes: the capitalist class, the managerial class, the small business class and the _____. (p. 206)
    a. large-business class
    b. executive class
    c. Marxian class
    d. working class

27. In examining the unequal distribution of income and wealth in the United States, the text notes that: (p. 209)
    a. the wealthiest 20 percent of households receive almost 50 percent of the total income "pie."
    b. the poorest 20 percent of households receive about 30 percent of the total income "pie."
    c. income inequalities between the rich and the poor narrowed greatly in 1999.
    d. differences in median income of married couples and female-headed households narrowed during the 1990's.

28. In the last two decades of the twentieth century, the gulf between the rich and the poor: (p. 209)
    a. greatly fluctuated.
    b. decreased dramatically.
    c. widened
    d. stayed the same

29. Which of the following is a major cause of poverty, according to the author of your textbook? (p. 219)
    a. low wages
    b. a flawed welfare system
    c. high employment rates
    d. the national shift from service to manufacturing jobs

30. According to the functionalist explanation of social inequality: (p. 220)
    a. all societies have important tasks that must be accomplished and certain positions that must be filled.
    b. the most important positions must be filled by the most qualified people.
    c. the most highly rewarded positions should be those that are functionally unique and on which other positions rely.
    d. all of the above.

TRUE- FALSE QUESTIONS

T F 1. According to Max Weber, lifestyle describes the extent to which persons within a particular layer of stratification have access to important scarce resources. (p. 194)

T F 2. People no longer believe in the "American Dream." (p.196)

T F 3. Both Karl Marx and Max Weber viewed class as an important determinant of social inequality. (p. 198)

T F 4. SES reflects the Weber multidimensional approach to determine social class. (p. 201)

T F 5. Most people in the U.S. identify themselves as the middle class. (p. 202)

T F 6. Of all the class categories, according to the Weberian Model, the one most shaped by formal education is the upper class. (p. 203)

T F 7. Accounting to the Weberian model of social class, an estimated thirty percent of the U.S. population is the working class (p. 204)

T F 8. The working poor account for about 10 percent of the U.S. population. (p. 204)

T F 9. About 3 to 5 percent of the U.S. population is the underclass. (p. 205)

T F 10. People of color have owned small businesses in the United States throughout U.S. history. (p. 207)

T F 11. The Marxian model suggests that the working class consists of about 50 percent of all employees of the U.S. (p. 208)

T F 12. Relative poverty exists when people do not have the means to secure the most basic necessities of life. (p. 216)

T F 13. About two third of all adults living in poverty are men. (p. 218)

T F 14. TANF, as a social welfare program, was replaced by AFDC. (p. 219)

T F 15. The Davis Moore thesis assumes that social stratification Results in meritocracy. (p. 220)

SOCIOLOGY IN OUR TIMES: DIVERSITY ISSUES

1. Do you believe in the American Dream? If so, what factors do you think will help you reach your dream? What factors do you think may limit your opportunities to reach that dream?
2. Why has education been a major factor in the upward mobility of Asian Americans, Latinos/as, and African Americans over the past fifty years? To what extent do you think education will help you achieve your own goals?
3. How do recent movies and television shows portray wealth and poverty? Are race, ethnicity and gender intertwined with these class depictions?
4. According to the text, the United States will become a better nation if it attempts to regain the American Dream by attacking poverty. Do you agree with this statement? Why or why not?

INTERNET ACTIVITIES

1. To read about welfare reform, go to the Urban Institute, a non-partisan economic and social policy research organization. Once there, look for the ***Focus on…*** box, and click on Welfare Reform to find current research on this topic.
   www.urban.org

2. The Bureau of Labor Statistics provides a wide range of economic data. Examine the Economy at a Glance section, which provides a regional and state-by-state breakdown of economic data. Then, read the section on Industry at a Glance, which provides the same kind of information on specific industries. Click on your state or the region in which you live to get current local information.
   www.bls.gov/eag/

3. Research Harvard University's Inequality and Social Policy Web Site Think Tank Links. Explore some of the ideas that people are talking and "thinking" about in regards to inequality and social policy.
   www.ksg.harvard.edu/inequality/Focus/links.htm

4. Use these websites to collect facts about poverty in the United States. Guide your students to use critical thinking skills to construct reasoned judgments about "who are the poor?"
   The U.S. Census Bureau Income Statistics
   www.census.gov/ftp/pub/hhes/www/income.html
   The U.S. Census Bureau Poverty Statistics
   www.census.gov/ftp/pub/hhes/www/poverty.html
   The 2001 Health and Human Services 2001 Poverty Guidelines
   http://aspe.hhs.gov/poverty/01poverty.htm

INFOTRAC EXERCISES

1. **Social Class**. Search for articles that address the plight of poor children. Bring a summary of the articles to class. Collectively construct a picture of the problems facing children in poverty.

2. Do a keyword search for the **American Dream**. What are the different ways that this term is used? What articles use the term to address social class issues? How often is this term used in a positive way? How has it become a negative symbol in our culture? Why do you think we use the word *dream* when we talk about our own social condition?

3. Look up **Karl Marx**. There are a number of references to his work The Communist Manifesto. You may want to actually read portions of this historic document.

4. Research five articles using InfoTrac. These articles must include all of the following terms: **Caste System**, **Class System**, **Slavery** and **Poverty**. State the journal, author, and title of each article. Once you complete this, write a brief synopsis on each of the 5 articles. Make certain that your paper meets your professor's requirements.

5. Discover the relationship between **social networks** and **occupational prestige**. An excellent source is the article, *Social Networks and Prestige Attainment: New Empirical Findings* (The American Journal of Economics and Sociology, Oct 1999).

STUDENT CLASS PROJECTS AND ACTIVITIES

1. Using theories, concepts, facts, and discussions found in this chapter, as well as from other sources, write a proposal for a long-term plan that would reduce income inequality in our nation. Write your proposal as if you were going to present it to Congress. Document your facts and figures and present compelling

evidence and information supporting your proposal. You must give sufficient goals and objectives of your proposal and provide proof that your plan is a feasible one. Type up your proposal and submit your paper at the appropriate time and in the style provided by your instructor.

2. Examine social stratification in either your own hometown, or the town/city in which your now reside. Conduct a driving route that will take you into a variety of neighborhoods that represent the six social classes as discussed on pages 280-283 of the text. You can probably secure census track data that will be particularly helpful for this project. Drive through the various neighborhoods recording the characteristics of the neighborhoods, the type of housing, probable lifestyles, the streets and the condition of the streets, the landscaping, any vehicles, and other signs that would indicate social class. Look for symbols of wealth or poverty, such as fences, recreational facilities, statues, flagpoles, water fountains, trees, and noise level, or the absence of such. Record anything unique or special about each neighborhood. Look for children in the neighborhoods and describe the children and their activities. Look for signs of conspicuous consumption and/or conspicuous waste. Submit a five-page paper describing each neighborhood using the above information as specific guidelines for the descriptions. In the writing of your project, focus on what you learned and how you felt about conducting this project. Follow any other specific guidelines provided by your instructor (if there are any) and submit your paper in the appropriate format at the appropriate time.

3. Research the historical creation of the "War on Poverty" programs. What book, written by sociologist Michael Harrington, served as a "reader" to the then President John F. Kennedy that documented the vast existence of U.S. poverty? What were some of that book's startling statistics? Which U. S. President declared war on poverty? When? Where? Why? How did the "name" of the programs—the War on Poverty originate? What was the country's first response to this call to fight poverty? What major programs were set up as a part of this "war?" Name and describe the programs. What eventually happened to the "War on Poverty?" What programs still exist? Write up and submit your paper, following any specific directions from your professor.

ANSWERS TO PRACTICE TESTS, CHAPTER 7

Answers to the Multiple Choice Questions

1. c An idea that often times gets taken for granted where the rich often times remove from the playing field (p. 193)
2. d All of the above apply (p. 193)
3. d Life chances are equated with opportunities (p. 194)
4. a The classic definition of social stratification (p. 194)
5. d As mentioned above, life chances refers to access to important societal resources. (p. 194)
6. b She has experienced intergenerational mobility (p. 194)
7. c Slavery still exists in various parts of the world even today. (pp 195-197)
8. d A caste system permanently determines people's status. (p. 198)

9.	b	An example of vertical, intergenerational mobility . (p. 198)
10.	c	According to Marx, class relationships involve inequality and exploitation (p. 199)
11.	a	Alienation is a feeling of powerlessness and estrangement from other people and from oneself. (p. 199)
12.	c	An egalitarian society is one in which class would not exist (p. 200)
13.	a	A classic example of symbolic interactionist perspective (p. 221)
14.	d	This is not one of Marx's analysis of class (p.p. 199-200)
15.	c	According to Weber, the definition of prestige. (p. 201)
16.	c	As identified by Max Weber (p. 201)
17.	b	45.8 million people in the United states are without health insurance (p. 211)
18.	b	According to Weber, the lower upper class has earned most of their money in their own lifetime. (p. 203)
19.	c	Inherited wealth is not one of the factors that qualifies people for the upper middle class (p. 203)
20.	a	Education is viewed by many as a means of attaining the American Dream (p. 203)
21.	a	Women in pink collar occupations are mainly classified in the working class. (p. 204
22.	b	According to the Weberian Model, these groups make up the middle class. (p. 203)
23.	d	The classic definition of the feminization of poverty (p. 201)
24.	b	This local school board position is not a component of wealth (p. 201)
25.	d	Often a class that becomes homeless while working (p. 205)
26.	d	This is the last of the four classes. (p. 206)
27.	a	The wealthiest 20% of U.S. households have half of the U.S's total income (p. 209)
28.	c	The gap between the rich and the poor has, in fact, widened. (p. 209)
29.	a	Low wages paid for many jobs is the major cause. (p. 219)
30.	d	This is the definite functionalist explanation for social inequality (p. 220)

Answers to True-False Questions

1.	False	According to Weber, **life chances** refer to the extent to which individuals have access to important societal resources. (p. 194)
2.	False	People still believe in the American dream. (p. 196)
3.	True	(p. 198)
4.	True	(p. 201)

5.	True	(p. 202)
6.	False	The upper middle class is the one most shaped by education. (p. 203)
7.	True	(p. 204)
8.	False	The working poor account for about 10 percent of the U.S. population. (p. 204)
9.	True	(p. 205)
10.	True	(p. 207)
11.	True	(p. 208)
12.	False	This is the definition of absolute poverty. (p. 216)
13.	False	About two-thirds of all adults living in poverty are women. (p.218)
14.	True	(p. 219)
15.	True	(p. 220)

CHAPTER 8
GLOBAL STRATIFICATION

BRIEF CHAPTER OUTLINE

Wealth and Poverty in Global Perspective

Problems in Studying Global Inequality

- The "Three Worlds" approach
- The Levels of Development Approach

Classification of Economies by Income

- Low-income economies
- Middle-income economies
- High-income economies

Measuring Global Wealth and Poverty

- Absolute, Relative, and Subjective Poverty
- The Gini Coefficient and Global Quality of Life Issues

Global Poverty and Human Development Issues

- Life expectancy
- Health
- Education and Literacy
- Persistent Gaps in Human Development

Theories of Global Inequality

- Development and Modernization Theory
- Dependency Theory
- World Systems Theory
- The New International Division of Labor Theory

Global Inequality in the Future

CHAPTER SUMMARY

Global stratification refers to the unequal distribution of wealth, power, and prestige on a global basis, resulting in people having vastly different lifestyles and life chances, both within and among the nations of the world. The social and economic gaps between the developed nations and the developing nations of the world are much more pronounced than they are in the United States. One approach to defining global stratification is through the "Three Worlds" approach. First world nations are the rich, industrialized countries having primarily capitalistic economies and democratic political systems. Second World nations are those countries having a moderate level of economic development and a moderate standard of living. Third World countries are the poorest countries with little or no industrialization, having lowest standards of living, shortest life expectancy, and highest mortality. Closely linked to the "three worlds" concept are the levels of development approach using terms such as developed nations, developing nations, less developed nations, and underdevelopment. The World Bank classifies nations into three economic categories: low-income economies, middle-income economies, and high-income economies. Using the gross domestic product is now a means of measuring wealth and power on a global basis. Global poverty is sometimes defined in terms of *absolute*, *relative*, and *subjective* poverty. The World Bank uses the Gini coefficient as its measure of income inequality. Using the Human Development

Index, the United Nations Development Program has established criteria for measuring the level of development in a country: life expectancy, education, and living standards. Because of social exclusion, a *fourth world* is developing; one where people are being systematically barred from access to positions that would enable their having an autonomous livelihood. Overall, the gap between the poorest nations and the middle-income nations has continued to widen. Social scientist use four primary theoretical perspectives to explain the persistence of global inequality: (1) development and **modernization theory**; (2) **dependency theory**; (3) world systems theory and (4) the new international division of labor theory. The future prospects of global inequality range from more to less optimistic predictions. Most analysts agree a nation of people can enjoy global prosperity only by ensuing that other people around the world have the opportunity to survive and thrive in their own surroundings.

LEARNING OBJECTIVES

After reading Chapter 8, you should be able to:

1. Define global stratification and explain how it contributes to economics inequality.

2. Identify and explain the two dimensions of global stratification according to Gunner Myrdal.

3. Define and describe the "three worlds" approach used to classify nations of the world.

4. Explain the levels of development approach used for describing global stratification.

5. Identify and explain the fourth world approach for describing global inequality.

6. Classify and describe nations of the world by the three economic categories.

7. Explain how poverty is defined on a global basis.

8. Identify and explain the use of the Gini coefficient.

9. Discuss global poverty and its effects upon human development.

10. Explain how and why slavery continues to exist in the global economy.

11. Compare and contrast three major theories of global inequality: development and modernization theory, dependency theory, and world systems theory.

12. Explain the new international division of labor theory.

13. Describe the future prospects of global inequality.

KEY TERMS (defined at page number shown and in glossary)

CHAPTER OUTLINE

I. WEALTH AND POVERTY IN GLOBAL PERSPECTIVE
   A. **Global stratification** refers to the unequal distribution of wealth, power, and prestige on a global basis.
   B. Global stratification results in people having vastly different life styles and life chance both within and among the nations of the world.
   C. The world is divided into unequal segments characterized by *high-income countries*, *middle-income countries*, and *low-income* countries; however, economic inequality is not the only dimension of global stratification; social inequality may result from factors such as discrimination based on race, ethnicity, gender, or religion.

II. PROBLEMS IN STUDYING GLOBAL INEQUALITY
   A. One of the problems is determining what terminology should be used to refer to the distribution of resources in various nations.
   B. After World War II, the "three worlds" approach was utilized to distinguish among nations based upon their economic development and their standard of living.
      1. First World nations consist of the rich, industrialized nations that primarily have capitalistic economic systems and democratic political systems.
      2. Second World nations consist of the countries with at least moderate level of economic development and a moderate standard of living.
      3. Third World nations consist of the poorest countries with little or no industrialization and the lowest standards of living, shortest life expectancies, and high mortality rate.
      4. Recently, the term *fourth world* is used to describe the "multiple black holes of social exclusion" of people in wide-range areas of the world, including those from sub-Saharan Africa to U.S. inner-city ghettos.
   C. The levels of development approach include concepts such as developed nations, developing nations, less developed nations, and underdevelopment.
      1. Leaders of the developed nations argue that economic development and growth is the primary way to solve poverty problems of the underdeveloped nations.
      2. This viewpoint requires that people in the lesser developed nations accept the beliefs and values of people in the developed nations.

3. If nations could increase the GNP, then social and economic inequality among their citizens could be reduced.
4. However, improving a country's GNP did not tend to reduce the poverty of the poorest people in that country and inequality increased even with greater economic development; some analysts blamed high rates of population growth in the underdeveloped nations.

III. CLASSIFICATION OF ECONOMIES BY INCOME

A. The World Bank classifies nations into three economic categories.
B. Low-income economies are nations that have a GNP per capita of $765 or less in 2003. About half of the world's population lives in the sixty-one low-income economies, with more women around the world tending to be more impoverished than men—a situation known as *the global feminization of poverty*
C. Middle-income economies are nations that have a GNP per capita of between $766 and $9,385 in 2003. About one-third of the world's population resides in the ninety-three nations with middle-income economies. These nations typically have a higher standard of living and export diverse goods and services, ranging from manufactured goods to raw materials and fuels.
D. High-income economies are found in fifty-four nations that have a GNI per capita of more than $9,385 in 2003 and continue to dominate the world economy, despite *capital flight* and *deindustrialization.*

IV. MEASURING GLOBAL WEALTH AND POVERTY

A. In recent years the United Nations and World Bank have begun to use the gross domestic product measurement—all the goods and services produced within a country's economy during a given year.
B. Global poverty is sometimes defined in terms of absolute, relative, and subjective poverty.
C. The Gina Coefficient and Global Quality of Life Issues
   1. The World Bank uses as its measure of income inequality the Gini coefficient, which ranges from zero (when everyone has the same income) to 100 (when one person receives all the income).
   2. Using this measure, the World Bank concluded that inequality increased in specific nations in 2003.

V. GLOBAL POVERTY AND HUMAN DEVELOPMENT ISSUES

A. Since the 1970s, the United Nations has focused on human development as a crucial factor in fighting poverty.
B. *Average life expectancy* has increased by about a third in the past three decades and is now more than 70 years in 87 countries; however, no country has reached a life expectancy for men of 80 years, but 19 countries have now reached an expectancy of 80 years or more for women.
C. *Health* is defined by the World Health Organization as a stage of complete physical, mental, and social well-being and not merely the absence of disease or infirmity. Many people in low-income nations suffer from infectious and other diseases. New diseases have recently emerged in countries all over the world,

while some middle-income countries are experiencing rapid growth in degenerative diseases such as cancer and coronary heart diseases.

D. *Education* is fundamental to increasing literacy.
   1. What is literacy and why is it important for human development? UNESCO defines a literate person as "someone who can, with understanding, both read and write a short, simple statement on their everyday life."
   2. The adult literacy rate in the low-income countries is about half of that of the high-income countries.

E. *Persistent gaps* in human development paint an overall dismal picture for the world's poorest people.
   1. The gap between the poorest nations and the middle-income nations has continued to widen.
   2. Although more women have paid employment than in the past, more and more women are still finding themselves in poverty because of increases in single person and single-parent households headed by women and low-wage work employment.

VI. THEORIES OF GLOBAL INEQUALITY

A. The most widely known theory is the development and **modernization theory**.
   1. This perspective links global inequality to different levels of economic development and suggests that low-income economies can move to middle- and high-income economies by achieving self-sustained economic growth.
   2. Along with intensive economic growth, the less developed nations can improve their standard of living with accompanying changes in people's beliefs, values, and attitudes toward work.

B. **Dependency theory** states that global poverty can partially be attributed to the exploitation of the low-income countries by the high-income countries.
   1. Dependency theory has been more often applied to the newly industrializing countries (NICs) of Latin America.
   2. Scholars examining the NICs of East Asia found that dependency theory had little or no relevance to economic growth and development in that part of the world.

C. World Systems theory suggests that under capitalism, a global system is held together by economic ties.
   1. The capitalist world-economy is a global system divided into a hierarchy of three major types of nations-core, semiperipheral, and peripheral.
   2. **Core nations** are dominant capitalist centers characterized by high levels of industrialization and urbanization.
   3. **Semiperipheral nations** are more developed than peripheral nations, but less developed than core nations.
   4. **Peripheral nations** are dependent on core nations for capital, have little or no industrialization (other than brought in by core nations), and have uneven patterns of urbanization.

D. According to the new international division of labor theory, commodity production is being split into fragments that can be assigned to whichever part of the world can provide the most profitable combination of capital and labor.

1. The global nature of these activities is referred to as *global commodity chains*, a complex pattern of international labor and production.
2. *Producer-driven* commodity chains describe industries in which transnational corporations play a central part in controlling the production process.
3. *Buyer-driven commodity chains* refers to industries wherein large retailers, brand-named merchandisers, and trading companies set up decentralized production network in various middle-and low-income counties.

VII. GLOBAL INEQUALITY IN THE FUTURE

A. Social scientists describe an optimistic or a pessimistic scenario for the future depending upon the theoretical framework they apply in studying global inequality.
   1. Some analysts highlight the human rights issues embedded in global inequality; others focus primarily on an economic framework.
   2. In the future, continued population growth, urbanization, and environmental degradation threaten even the meager living conditions of those residing in high-income countries; the quality of life diminishes as natural resources are depleted.
   3. A more optimistic scenario suggests that with modern technology and worldwide economic growth, it might be possible to reduce absolute poverty and to increase people's opportunities, ensuring that people all over the world have the opportunity to survive and thrive in their own surroundings.

ANALYZING AND UNDERSTANDING THE BOXES

After reading the chapter and studying the outline, re-read the boxes and write down key points and possible questions for class discussion.

Sociology and Everyday Life: How Much Do You Know About Global Wealth and Poverty?

Key Points:

Discussion Questions:

1.

2.

3.

Sociology and Social Policy: Should we in the United States do something about Child Labor in other Nations?

Key Points:

Discussion Questions:

1.

2.

3.

You Can Make a Difference: Global Networking to Reduce World Hunger and Poverty

Key Points:

Discussion Questions:

1.

2.

3.

PRACTICE TESTS

MULTIPLE CHOICE QUESTIONS

Select the response that best answers the question or complete the statement:

1. The unequal distribution of wealth, power, and prestige on a global basis is referred to as: (p. 228)
   a. global stratification.
   b. global layering.
   c. global distinction.
   d. global accumulation.

2. The income gap between the richest and the poorest 20 percent of the world population: (p. 228)
   a. continues to widen
   b is greater in urban than in rural areas
   c. has significantly decreased in the last decade
   d. is slowly beginning to decline

3. The _____approach was introduced by social analysts to distinguish among nations on the basis of their levels of economic development and the standard of living of their citizens. (p. 231)
   a. levels of development approach
   b. three worlds approach
   c. classification of economies approach
   d. global distinction approach

4. China and Cuba are classified as ____nations. (p. 231)
   a. "developed world"
   b. "first world"
   c. "second world"
   d. "third world

5. First world nations are best represented by the countries of: (p. 231)
   a. Japan, the United States, China.
   b. New Zealand, the United States, Great Britain.
   c. Canada, China, Korea.
   d. Russia, Australia, Japan.

6. The ____plan, named after a U. S. Secretary of State provided massive sums of money in direct aid and loans to re-build countries destroyed during World War II. (p. 232)
   a. Albright Plan
   b. Powell Plan
   c. Marshall Plan
   d. Rice plan

7. The term _____ refers to all the goods and services produced in a country in a given year. (p. 232)
   a. WHO
   b. GDP
   c. NIC
   d. GNI

8. Low-income economies are primarily found in _____ nations, where half of the world's population resides. (pp. 232-233)
   a. Asian and South American
   b. Asian and African
   c. Eastern European and African
   d. African and South American

9. The situation wherein women around the world tend to be more impoverished than men is referred to as: (p. 233)
   a. global feminization stratification.
   b. global feminization.
   c. global feminization of poverty.
   d. global feminization of labor.

10. Lower-middle income economies include the nations of: (p. 233)
    a. Russia, Romania, Kazakhstan.
    b. South Korea, the Ukraine, Mexico.
    c. Mexico, Guatemala, Russia.
    d. Mozambique, Bosnia, Turkey.

11. High-income economies are found in _____ nations. (p. 235)
    a. 54
    b. 15
    c. 38
    d. 25

12. The movement of jobs and economic resources from one nation to another is defined as: (p. 235)
    a. capital flight.
    b. deindustrialization.
    c. transnational flight.
    d. economic mobility

13. The closing of plants and factories because of their obsolescence or employment of cheaper workers in other nations is known as: (p. 235)
    a. capital flight.
    b. capital destabilization.
    c. deindustrialization.
    d. reindustrialization.

14. The World Bank uses the term _____ to indicate all of the goods and services produced within a country's economy during a given year. (p.235)
    a. GDP
    b. ENP
    c. GPA
    d. GPO

15. _____ poverty is a condition in which people do not have the means to secure the most basic necessities of life. (p. 236)
    a. subjective poverty
    b. relative poverty
    c. absolute poverty
    d. standard poverty

16. The World Bank uses the term _____ as its measure of income inequality. (p. 236)
    a. GNP
    b. Gini coefficient
    c. relative poverty
    d. Mrydal coefficient

17. According to the _____ theory, the capitalist world-economy is a global system divided into a hierarchy of three major types of nations. (p. 244)
    a. dependency
    b. development and modernization
    c. new international division of labor
    d. world systems

18. What process usually brings with it a higher standard of living in a nation and some degree of social mobility for individual participants? (p. 239)
    a. Industrialization
    b. Capitalization
    c. Urbanization
    d. Welfare Socialism

19. The term for the stage in Rostow's modernization theory wherein there is a belief in individualism, competition, and achievement is the: (p. 242)
    a. traditional stage
    b. historical stage.
    c. take-off stage
    d. technological maturity.

20. Which of the following is NOT a reasonable criticism of the modernization theory? (p. 242)
    a. The theory ignores the exploitation of low-income to high-income economies.
    b. The theory tends to be Eurocentric
    c. Transnational corporations are now the major decision makers making the rules and regulations of any one nation irrelevant.
    d. The theory blames the low-income countries for their own condition.

21. Which of the following is supported by dependency theory? (p. 244)
    a. High-income countries are the models for the economic development of low-income countries.
    b. Global capitalization will narrow the gap between the low-income and high-income countries.
    c. Low-income countries can develop by achieving self-sustained economic growth.
    d. global poverty historically stems from the exploitation of low-income by high income.

22. What contemporary theorist best represents world systems theory? (p.244)
    a. Immanual Wallerstein
    b. Walt Rostow
    c. Gunner Myrdal
    d. Gary Gereffi.

23. According to Wallerstein, what term represents the high-income countries of the world? (p. 245)
    a. Core.
    b. Semiperiphery
    c. High-tech
    d. Mixed-tech

24. According to the _______theory, commodity production is split into fragments, each of which can be moved to the part of the world that can provide the best combination of capital and labor. (p. 246)
    a. world systems
    b. new international division of labor
    c. dependency
    d. modernization

25. The term global commodity chains is most closely associate with _________ (p.246)
    a. dependency
    b. world systems
    c. new international division of labor
    d. modernization

26. ____are types of chains most common in labor-intensive consumer goods industries, such as toys and footwear. (p. 247)
    a. Producer-driven commodity chains
    b. Buyer-driven commodity chains
    c. Individually-driven commodity chains
    d. Transnational-driven commodity chains

27. Which nations are at the periphery of the world economy according to world system theory? (p. 245)
    a. High-income countries
    b. Middle-income countries
    c. Low-income countries
    d. Mixed-income countries

28. Dependency theory most closely approximates a/an ________approach (p. 244)
    a. interactionist
    b. conflict
    c. functionalist
    d. evolutionary

29. Which of the following does not represent semiperipheral nations according to world systems theory? (p. 245)
    a. Mexico
    b. India
    c. Brazil
    d. the Caribbean

30. Maquilador plants, wherein low-wage workers assemble products that are then brought into the United States, are found in what country? (p. 245)
    a. Haiti
    b. Cuba
    c. the Caribbean
    d. Mexico

TRUE-FALSE QUESTIONS

T F 1. In utilizing the "three worlds" approach for studying global inequality, North Korea is an example of a Third World Nation. (p. 231)

T F 2. Ideas regarding underdevelopment were popularized by U.S. President Harry S. Truman. (p. 232)

T F 3. According to the development approach in studying global inequality, an increase in the standard of living of a century meant that a nation was moving toward economic development. (p.232)

T F 4. High rates of population growth taking place in the underdeveloped nations have led to an increase in greater economic development of those countries. (p.232)

T F 5. About half of the world's population lives in the 61 low-income economies, according to the World Bank classification (p. 232)

T F 6. Today, the World Bank classifies nations into three economic categories: low, middle, and high. (p. 232)

T F 7. According to the classification of economies by income, many countries referred to as "middle income" have very few of their population living in poverty. (p.234)

T F 8. The World Bank has defined absolute poverty as living on less that a dollar a day. (p. 236)

T F 9. Subjective poverty is defined as a condition in which people do not have the means to secure the most basic necessities.(p. 236)

T F 10. Since the 1970s, the United Nations has more actively focused on human development, as measured by the HDI as a crucial factor for fighting poverty. (p.236)

T F 11. The numbers of people dying worldwide from hunger-related diseases is the equivalent of 200 jumbo jet crashes per day with no survivors.(p.238)

T F 12. One major cause of shorter life expectancy in low-income nations is the high rate of infant mortality.(P.238)

T F 13. About 17 million people die each year from diarrhea, malaria, tuberculosis, and other infectious and parasitic illnesses. (p.236)

T F 14. According to UNESCO, women in low-income countries comprise about two- thirds of those who are illiterate.(p.239)

T F 15. Modernization theory most closely resembles the Functionalist Perspective. (p. 243)

STUDENT CLASS PROJECTS AND ACTIVITIES

1. Select one present independent country (or republic) that used to be a republic that comprised the former Union of Soviet Republics (before September, 1991): (1) Armenia; (2) Azerbaijan; (3) Belorussia; (4) ; (5) Georgia; (6) Kasakhstan; (7) Kirghizia; (8) Latvia; (9) Lithuania; (10) Moldavia; (11) Russian S.F.S.R.; (12) Tadshikistan; (13) Turkmenistan; (14) Ukraine; and (15) Uzbekistan. Or select one independent country that used to be part of a whole (such as Bosnia or Croatia or the former country of Yugoslavia). You are to prepare a research report of a minimum of five typewritten pages about their respective republic, or country. Your reports must include: (1) the population: (2) ethnic breakdown of various population groups; (3) date that the country first became a part of the whole and the date that the country became independent from the whole; (4) brief historical background of the country; (5) size (in square miles); (6) major goods, service, and raw products produced; (7) various religious practices; (8) language that is spoken; (9) typical lifestyle of the people; (10) the social and economic status of the people either within their own group or in comparison to an outside group, (11) political sentiments of the people, i.e., is there a strong sense of

nationalism; (12) any other information that enhances the understanding of this country or republic. Submit your paper at the appropriate time, following the instructions of your instructor.

2. Drawing upon World Systems Theory most closely associated with the work of Immanuel Wallerstein, select three countries that are members of ASEAN (Association of South East Asian Nations) who best represent the three major types of nations in the hierarchical position of nations in the global economy. Prepare a research report on the economic conditions of each of the three you selected, including information such as (1) the status of the major infrastructures of each country, such as their transportation systems, their power plants, their telecommunications systems; (2) the status of their financial capital and (3) the status of their human capital. i.e., the human skills needed to handle. Provide substantial explanations of why and how you selected the countries representing each specific category. Include an analysis of the resources and obstacles characterizing each specific country. Are there any global cities (or command posts) in any of the countries that you selected? Provide any other information that supports you selections.

3. In the later part of the twentieth century, countries in various parts of the globe began to form economic, political and/or social alliances. Research past and present alliances, noting the purpose of the alliance, the status of that alliance today, countries that are part of the alliance, and possible or probable future alliances. Some possibilities of the past, present, and future are (1) Atlantic Rim; (2) Asian Pacific (Pacific Rim); (3) the Group of Seven; (4) the Group of 77; (5) NAFTA; (6) Hemispheric Free Trade Area (HFTA); (7) Asia Pacific Economic Cooperation (APEC); (8) the EU (European Union); (9) North Atlantic Free Trade Area; (10) Mediterranean Rim; (11) NATO; and (12) OPEC. Research those alliances that were created during the years of this twenty-first century. Write up your paper in the specific form requested by your professor and submit it on the appropriate date.

4. Research the lives of the world's ten richest people, according to the latest Forbes magazine poll. (1) Provide a minimum one-paragraph biographical sketch on each of the "top ten." (2) Explain how each of them obtained their wealth. (3) What countries do they represent? (4) Who did they replace from the previous year's poll? (5) What are some similarities and differences among the top ten? Submit your paper at the appropriate time, following the instruction of your professor.

5. The world faces no shortage of problems—or of suggested ideas to solve them, such as the world population problems, energy problems, diseases, water and wealth, world poverty, as well as the sheer survival of animal, plant, and human species. According to the Special Issue of **Scientific American, September, 2005**, the world of nations is facing a unique turning point in the next twenty years, creating a "Crossroads for Planet Earth." This special edition puts forth a comprehensive plan for a bright future for planet earth beyond 2050, responding

to these problems. Research each topic in this special edition, or of a similar resource. Identify each specific world problem mentioned above, and provide a probable and possible solution to each. Write and submit your paper/project following specific instructions of your professor.

INTERNET ACTIVITIES

1. Visit the United Nations Economic and Social Development Web Site. This site has links to information about crime, trade, drugs, population, etc. from a global perspective.
   www.un.org/esa/
2. Another organization to visit is the World Bank Group. To read about poverty, click on Poverty Reduction Strategy Papers. What are some of the proposed ways to reduce or end poverty?
   www.worldbank.org
3. UNESCO stands for the United Nations Educational, Scientific, and Cultural Organization. What is the mission of UNESCO? To read a quick summary of important facts, click on *Did You Know*? This list of quick facts addresses literacy, primary education and childhood development, and who finances education. How will reducing global illiteracy help the United States, and other countries?
   www.unesco.org
4. A site devoted to international issues of women's health is the International Women's Health Coalition. Read about their population policies and efforts to increase AIDS awareness for women around the world.
   www.iwhc.org/
5. From these sites you can take a surfing trip to gain facts about stratification, poverty and conflict. To generate a deeper level of interest bring facts and resources to class. A typed up "fact sheet" can be turned in for credit.
   www.globalissues.org/TradeRelated/Poverty.asp
6. Visit this site to read an essay entitled "Global Poverty in the Late 20th Century" by Michel Chossudovsky, Professor of Economics at the University of Ottawa. Write three critical thinking questions based on their reading of the essay.
   www.heise.de/tp/english/special/eco/6099/1.html

INFOTRAC EXERCISES

1. Examine **Human Rights Violations** by searching for articles that address the human rights violations in Serbia, Russia, China, Saudi Arabia, and Sierra Leone to name a few.

2. Type in under the subject **Gini coefficient** and see how this measure of income inequality is used in actual research articles.

3. There are a number of articles under the subject **Women and Poverty**. Find articles related to global poverty. What cultural values are related in some of these articles?

4. There are a number of subdivisions under the subject search for **Developing Countries**. Under the subdivision Environmental Aspects students can learn about the interrelationship between the social structure and the physical environment.

ANSWERS TO PRACTICE TESTS FOR CHAPTER 8

Answers to Multiple Choice Questions

1.	a	Global stratification is the unequal global distribution of wealth, power, and prestige (p.228)
2.	a	The income gap between rich and poor continues to widen. (p.228)
3.	b	Three worlds approach is used to distinguish a nation's economic development and it's citizens standard of living. (p.231)
4.	c	Second world nations are countries that have a moderate level of economic development and moderate levels for standard of living. (p.231)
5.	b	First world nations are rich, industrial nations that usually have democratic political systems and a capitalist economy (p. 231)
6.	c	The Marshall Plan provided loans and aid to post WWII Europe. (p.232)
7.	d	GNI stands for gross national income. The GNI measures the total amount of goods and services produced a country annually. (p. 232)
8.	b	Generally, Asian and African nations tend to be low-income nations. (pp. 232-233)
9.	c	Global feminization of poverty: throughout many nations, women tend to be more impoverished than men. (p.233)
10.	a	These nations suffered economic crashes in the late 1980's and early 1990's. Since that time, their economies have yet to recover. (p. 233)
11.	d	the number of nations worldwide who are considered to have high-incomes. (p. 235)
12.	a	Moving economic resources and jobs to a new country. (p. 235)
13.	c	Many American companies are currently closing plants and factories and moving jobs overseas. (p. 235)
14.	a	GDP or gross domestic product is all the goods and services produced within a country's economy during a given year. (p. 235)
15.	c	Absolute poverty means that a person doesn't have the means to get the basic things in life. (p. 236)
16.	b	Gini Coefficient is used to measure income inequality. (p. 236)
17.	d	World systems theory is a theory that divides the world into three types of nations. 1. core 2. semiperipheral 3.peripheral. (p. 244)
18.	a	There becomes greater opportunities (p. 239)
19.	c	This is the beginning (p. 242)
20.	a	Explotation is a strong criticism. (p.242)
21.	d	Exploitation is a key factor for dependency theorists. (p. 244)
22.	a	He best represents world systems theory. (p.244)

23. a Core. (p. 237)
24. b It is taking Durkheimian theory and globalizing it (p.238)
25. c See #24 (p.246)
26. b Supply and demand. (p. 247)
27. c Low income countries (p.245)
28 b With exploitation, it fits well with conflict. (p. 244)
29. d The Carribean (p. 245)
30. d Mexico (p. 245)

Answers to True-False Questions:

1. False North Korea is an example of a second world nation. (p. 231)

2. True (p.232)

3. True (p.232)

4. False High rates of population grown in underdeveloped nations have not led to an increase in economic development. (p.232)

5. True (p. 232)

6. True (p. 232)

7. False Many "middle income" countries have large numbers of people living in poverty. (p234)

8. True (p. 236)

9. False Absolute poverty is defined as a condition in which people do not have the means to secure the most basic necessities. (p.236)

10. True (p. 236)

11. False The number of people dying is equivalent to 300 jumbo jet crashes. (p.238)

12. True (p. 238)

13. True (p.238)

14. True (p.239)

15. True (p.243)

CHAPTER 9
RACE AND ETHNICITY

BRIEF CHAPTER OUTLINE

Race and Ethnicity
- Social Significance of Race and Ethnicity
- Race Classifications and the Meaning of Race
- Dominant and Subordinate Groups

Prejudice
- Stereotypes
- Racism
- Theories of Prejudice

Discrimination

Sociological Perspectives on Race and Ethnic Relations
- Symbolic Interactionist Perspectives
- Functionalist Perspectives
- Conflict Perspectives
- An Alternative Perspective: Critical Race Theory

Racial and Ethnic Groups in the United States
- Native Americans
- White Anglo-Saxon Protestants (British Americans)
- African Americans
- White Ethnic Americans
- Asian Americans
- Latinos/as (Hispanic Americans)
- Middle Eastern Americans

Global Racial and Ethnic Inequality in the Future
- Worldwide Racial and Ethnic Struggles
- Growing Racial and Ethnic Diversity in the United States

CHAPTER SUMMARY

Issues of race and ethnicity permeate all levels of interaction in the United States. A **race** is a category of people who have been singled out as inferior or superior, often on the basis of physical characteristics such as skin color, hair texture, and eye shape. By contrast, an **ethnic group** is a collection of people distinguished, by others or by themselves, primarily on the basis of cultural or nationality characteristics. Race and ethnicity are ingrained in our consciousness and often form the basis of hierarchical ranking and determine who gets what resources. A **majority** (or **dominant**) **group** is one that is advantaged and has superior resources and rights in a society while a **minority** (or **subordinate**) **group** is one whose members, because of physical or cultural characteristics, are disadvantaged and subjected to unequal treatment by the dominant group and who regard themselves as objects of collective discrimination. **Prejudice** is a negative attitude based on faulty generalizations about the members of selected racial and ethnic groups. **Discrimination** - actions or practices of dominant-group members that have a harmful impact on members of a subordinate group, may be either **individual** or **institutional discrimination** which involves day-to-day practices of organizations and institutions that have a harmful impact on members of subordinate groups. According to the symbolic interactionist contact hypothesis, increased contact between people from divergent groups should lead to favorable attitudes and behavior when a specific set of criteria are met. Two functionalist perspectives, **assimilation** and **ethnic pluralism,** focus on how members of subordinate groups become a part of the mainstream. Alternately, conflict theories analyze economic stratification and access to power in race and ethnic relations: caste and class perspectives, **internal colonialism**, **split labor market** theory, gendered racism, and racial formation theory. Critical race theory derives its foundation from the U.S. civil rights tradition and the writings of several civil rights leaders. The unique experiences of Native Americans, White Anglo-Saxon Protestants/British Americans, African Americans, White Ethnics, Asian Americans, Latinos/as (Hispanic Americans), and Middle Easterners are discussed, and the increasing racial-ethnic diversity of the United States is examined. Globally, many racial and ethnic groups seek self-determination, creating ethnic wars in some areas. In the future, it is hoped that the superpower nations with the aid of the United Nations will suppress ethnic violence.

LEARNING OBJECTIVES

After reading Chapter 9, you should be able to:

1. Define race and ethnic group and explain their social significance.

2. Explain the sociological usage of majority group and minority group and note why these terms may be misleading.

3. Discuss prejudice and explain the major theories of prejudice.

4. Discuss discrimination and distinguish between individual and institutional discrimination.

5. Describe symbolic interactionist perspectives on racial and ethnic relations.

6. Distinguish between assimilation and ethnic pluralism and explain why both are functionalist perspectives on racial and ethnic relations.

7. Explain the key assumptions of conflict perspectives on racial and ethnic relations and note the group(s) to which each applies.

8. Know the major premises of critical race theory.

9. Trace the intergroup relationships of racial and ethnic groups in the United States.

10. Explain how the experiences of Native Americans have been different from those of other racial and ethnic groups in the United States.

11. Describe how the African American experience in the United States has been unique when compared with other groups.

12. Compare and contrast the experiences of Chinese Americans, Japanese Americans, Korean Americans, Filipino Americans, and Indochinese Americans in the United States.

13. Describe the experiences of Mexican Americans, Puerto Ricans, and Cuban Americans in the United States.

14. Define the major groups that comprise Middle Easterners living in the United States.

15. Discuss racial and ethnic struggles from a global perspective.

KEY TERMS (defined at page number shown and in glossary)

assimilation 263
authoritarian personality 260
discrimination 260
dominant group 258
ethnic group 255
ethnic pluralism 263
gendered racism 267
genocide 260
individual discrimination 261
institutional discrimination 261
internal colonialism 267
prejudice 258
race 254
racism 259
scapegoat 260
segregation 264
split labor market 267
stereotypes 258
subordinate group 258
theory of racial formation 268

CHAPTER OUTLINE

I. RACE AND ETHNICITY
   A. The social significance that people accord to race is more significant than any biological differences that might exist among people who are placed in arbitrary categories.
   B. A **race** is a category of people who have been singled out as inferior or superior, often on the basis of physical characteristics such as skin color, hair texture, and eye shape.
   C. An **ethnic group** is a collection of people distinguished, by others or by themselves, primarily on the basis of cultural or nationality characteristics.
   D. Race and ethnicity take on great social significance because how people act in regards to these terms drastically affects people's lives and are bases of hierarchical ranking in society; the dominant group holds power over other (subordinate) ethnic groups.
   E. Racial classifications in the U.S. census mirror how the meaning of race has continued to change over the past century in the U.S.

F. A **majority** (or **dominant**) **group** is one that is advantaged and has superior resources and rights in a society; a **minority** (or **subordinate**) **group** is one whose members, because of physical or cultural characteristics, are disadvantaged and subjected to unequal treatment by the dominant group and who regard themselves as objects of collective discrimination.

II. PREJUDICE

A. **Prejudice** is a negative attitude based on faulty generalizations about members of selected racial and ethnic groups. Prejudice is often based on stereotypes.

B. **Stereotypes** are overgeneralizations about the appearance, behavior, or other characteristics of all members of a category.

C. **Racism** is a set of attitudes, beliefs, and practices that is used to justify the superior treatment of one racial or ethnic group and the inferior treatment of another racial or ethnic group.

D. Theories of prejudice include the frustration aggression hypothesis, which states that people who are frustrated in their efforts to achieve a highly desired goal will respond with a pattern of aggression toward a **scapegoat** -- a person or group that is incapable of offering resistance to the hostility or aggression of others. The theory of the **authoritarian personality**, one which is characterized by excessive conformity, submissiveness to authority, intolerance, insecurity, a high level of superstition, and rigid, stereotypic thinking, characterizes a highly prejudiced person, according to Adorno and others.

III. DISCRIMINATION

A. **Discrimination** is defined as actions or practices of dominant group members that have a harmful impact on members of a subordinate group.

B. Robert Merton identified four combinations of attitudes and responses:

1. Unprejudiced nondiscriminator persons who are not personally prejudiced and do not discriminate against others;
2. Unprejudiced discriminators persons who may have no personal prejudice but still engage in discriminatory behavior because of peer group prejudice or economic, political, or social interests;
3. Prejudiced nondiscriminatory persons who hold personal prejudices but do not discriminate due to peer pressure, legal demands, or a desire for profits; and
4. Prejudiced discriminators persons who hold personal prejudices and actively discriminate against others.

C. Discriminatory actions vary in severity from the use of derogatory labels to violence against individuals and groups.

1. **Genocide** is the deliberate, systematic killing of an entire people or nation.
2. More recently, the term "ethnic cleansing" has been used to define a policy of "cleansing" geographic areas (such as in Bosnia-Herzegovina) by forcing persons of other races or religions to flee or die.

D. Discrimination also varies in how it is carried out.

1. **Individual discrimination** consists of one-on-one acts by members of the dominant group that harm members of the subordinate group or their property.

2. **Institutional discrimination** is the day-to-day practices of organizations and institutions that have a harmful impact on members of subordinate groups.

IV. SOCIOLOGICAL PERSPECTIVES ON RACE AND ETHNIC RELATIONS
- A. Symbolic Interactionist Perspectives
    1. The *contact hypothesis* suggests that contact between people from divergent groups should lead to favorable attitudes and behavior when a specific set of criteria is met.
    2. However, scholars have found that increasing contact may have little or no effect on existing prejudices.
- B. Functionalist Perspectives
    1. **Assimilation** is a process by which members of subordinate racial and ethnic groups become absorbed into the dominant culture.
    2. **Ethnic pluralism** is the coexistence of a variety of distinct racial and ethnic groups within one society.
    3. **Segregation** is the spatial and social separation of categories of people by race, ethnicity, class, gender, and/or religion.
- C. Conflict Perspectives
    1. The *caste perspective* views racial and ethnic inequality as a permanent feature of U.S. society.
    2. *Class perspectives* emphasize the role of the capitalist class in racial exploitation.
    3. **Internal colonialism** occurs when members of a racial or ethnic group are conquered or colonized and forcibly placed under the economic and political control of the dominant group.
    4. **Split labor market** refers to the division of the economy into two areas of employment, a primary sector composed of higher paid (usually dominant-group) workers in more secure jobs, and a secondary sector comprised of lower paid (often subordinate-group) workers in jobs with little security and frequently hazardous working conditions.
    5. *Gendered racism* refers to the interactive effect of racism and sexism in the exploitation of women of color.
    6. The *theory of racial formation* states that actions of the government substantially define racial and ethnic relations in the United States.
- D. An alternative perspective: Critical Race theory
    1. One premise is that racism is so ingrained in the U.S. society that it appears to be ordinary and natural to many people.
    2. *Interest convergence* is a crucial factor in bringing about social change.
    3. Formal equality under the law does not necessarily equate to actual equality in society.

V. RACIAL AND ETHNIC GROUPS IN THE UNITED STATES
- A. Native Americans
    1. Historically, Native Americans experienced the following kinds of treatment in the United States:

a. Genocide
b. Forced Migration
c. Forced Assimilation

2. Today, about 2.5 million Native Americans live in the United States (primarily in the southwest), and about one-third live on reservations.
3. Native Americans are the most disadvantaged racial or ethnic group in the United States in terms of income, employment, housing, nutrition, and health (especially among individuals living on reservations).

B. White Anglo-Saxon Protestants (British Americans)

1. Although many English settlers initially were indentured servants or sent here as prisoners, they quickly emerged as the dominant group, creating a core culture to which all other groups were expected to adapt.
2. Like other racial and ethnic groups, British Americans are not all alike; social class and gender affect their life chances and opportunities.

C. African Americans

1. Slavery was rationalized by stereotyping African Americans as inferior and childlike; however, some slaves and whites engaged in active resistance that eventually led to the abolition of slavery.
2. Through informal practices in the North and Jim Crow laws in the South, African Americans experienced segregation in housing, employment, education, and all public accommodations.
3. *Lynching*, a killing carried out by a group of vigilantes seeking revenge for an actual or imagined crime by the victim, was used by whites to intimidate African Americans into staying "in their place."
4. During World Wars I and II, African Americans were a vital source of labor in war production industries; however, racial discrimination continued both on and off the job.
   a. After African Americans began to demand sweeping societal changes in the 1950s, through the civil rights movement utilizing *civil disobedience*, racial segregation slowly was outlawed by the courts and the federal government.
   b. Civil rights legislation attempted to do away with discrimination in education, housing, employment, and health care.
5. Today, African Americans make up about 13 percent of the U.S. population; many have made significant gains in education, employment, and income in the past three decades; however, other African Americans have not fared so well; for example, the African American unemployment rate remains twice as high as that of whites, and young people are far less likely to have a home computer.

D. The term *white ethnics* is applied to a wide diversity of immigrants who trace their origins from European countries other than England: Ireland, Poland, Italy, Greece, Germany, Yugoslavia, Russia and other former Soviet republics.

1. Both Irish Americans and Italian Americans were subjected to institutional discrimination in employment.
2. Jews were victims of *anti-Semitism*.

3. Sports provided paths to assimilation for many white ethnics: the earliest non Anglo-Saxon Protestant football players were of Irish, Italian, and Jewish ancestry.

E. Asian Americans

1. Chinese Americans
   a. The initial wave of Chinese immigration occurred between 1850 and 1880 when Chinese men came to the United States seeking gold in California and jobs constructing the transcontinental railroads.
   b. Chinese Americans were subjected to extreme prejudice and stereotyping; the Chinese Exclusion Act of 1882 was passed because white workers feared for their jobs; the law remained until World War II.
   c. In the 1960s, the second and largest wave of Chinese immigration came from Hong Kong and Taiwan.
   d. Today, as a group, they have enjoyed considerable upward mobility, however many Chinese Americans live in poverty in urban Chinatowns.
2. Japanese Americans
   a. The earliest Japanese immigrants primarily were men who worked on sugar plantations in the Hawaiian Islands in the 1860s; the immigration of Japanese men was curbed in 1908; however, Japanese women were permitted to enter the U.S. for several more years because of the shortage of women.
   b. Internment: During World War II, when the United States was at war with Japan, nearly 120,000 Japanese Americans were placed in internment camps because they were seen as a security threat; many Japanese Americans lost all that they owned during the internment.
   c. In spite of the extreme hardship faced as a result of the loss of their businesses and homes during World War II, many Japanese Americans have been very successful; their median income is more than 30 percent above the national average.
3. Korean Americans
   a. The first wave of Korean immigrants were male workers who arrived in Hawaii between 1903 and 1910; the second wave came to the mainland following the Korean War in 1954 (e.g., the wives of servicemen, and Korean children who had lost their parents in the war); and the third wave arrived after the Immigration Act of 1965 permitted well educated professionals to migrate to the U.S.
   b. Korean Americans have helped each other open small businesses by pooling money through the *key*- an association that grants members money on a rotating basis to gain access to more capital.
   c. Ongoing discord has existed between Korean Americans and African Americans in New York and California.

4. Filipino Americans
   a. Most of the first Filipino immigrants were men who were employed in agriculture; following the Immigration Act of 1965, Filipino physicians,

nurses, technical workers, and other professionals moved in large numbers to the U.S. mainland.

b. Unlike other Asian Americans, most Filipinos have not had the startup capital necessary to open their own businesses, and workers generally have been employed in the low wage sector of the dual labor market.

c. However, average household incomes of Filipino American families are high because about 75 percent of the women are employed and nearly half have a four-year college degree.

5. Indochinese Americans

a. Most Indochinese Americans (including people from Vietnam, Cambodia, Thailand, and Laos) have come to the U.S. in the past three decades.

b. Vietnamese refugees who had the resources to flee at the beginning of the Vietnam War were the first to arrive. Next came Cambodians and lowland Laotians, referred to as "boat people" by the media.

c. Today, most Indochinese Americans are foreign born; about half live in western states, especially California. Even though most Indochinese immigrants spoke no English when they arrived in this country, some of their children have done very well in school and have been stereotyped as "brains".

F. Latinos/as (Hispanic Americans)

1. The terms *Latino, Latina*, and *Hispanic* are used interchangeably to refer to people who trace their origins to Spanish-speaking Latin America and the Iberian peninsula; the label *Hispanic* was first used by the U.S. government to designate people of Latin American and Spanish descent living in the United States; the term has not been fully accepted.

2. Mexican Americans or Chicanos/as have experienced disproportionate poverty as a result of internal colonialism.

a. More recently, Mexican Americans have been seen as cheap labor at the same time that they have been stereotyped as lazy.

b. When anti-immigration sentiments are running high, Mexican Americans often are the objects of discrimination.

c. Today, the families of many Mexican Americans have lived in the United States for four or five generations and have made significant contributions in many areas.

3. When Puerto Rico became a possession of the United States in 1917, Puerto Ricans acquired U.S. citizenship and the right to move freely to and from the mainland; while living conditions and increased professionalism have improved substantially for some, others have continued to live in poverty in Spanish Harlem and other barrios.

4. Cuban Americans have fared somewhat better than other Latinos; early waves of Cuban immigrants were affluent business and professional people; the second wave of Cuban Americans fared worse; and more recent arrivals have developed their own ethnic and economic enclaves in cities such as Miami.

G. Middle Easterner Americans
  1. Since 1970, many immigrants have arrived in the United States from countries located in the "Middle East" such as Egypt, Syria, Lebanon, Iran, Iraq, Kuwait, and Jordan.
  2. Middle Eastern Americans speak a variety of languages and have diverse religious backgrounds, such as: Muslim, Coptic Christian, or Melkite Catholic.
  3. While some are from working class families, the Lebanese Americans, Syrian Americans, Iranian Americans, and Kuwati Americans primarily come from middle class backgrounds.

VI. GLOBAL RACIAL AND ETHNIC INEQUALITY IN THE FUTURE

A. Worldwide Racial and Ethnic Struggles
  1. The cost of self-determination the right to choose one's own way of life often is the loss of life and property in ethnic warfare (e.g., Bosnia-Herzegovina, Croatia, Spain, Britain, Romania, Russia, Moldova, Georgia, the Middle East, Africa, Asia, and Latin America).
  2. In the twenty-first century, the struggle between the Israeli government and various Palestinian factions over the future and borders of Palestine continues to make headlines.

B. Growing Racial and Ethnic Diversity in the United States
  1. Racial and ethnic diversity is increasing in the United States: African Americans, Latinos/as, Asian Americans, and Native Americans make up about one-fourth of the U.S. population; by 2056, the roots of the average U.S. resident will be Africa, Asia, Hispanic countries, the Pacific Islands, or Arabia not white Europe.
  2. Interethnic tensions may become more overt between whites and people of color; people may continue to employ *sincere fictions* personal beliefs that are a reflection of larger societal mythologies, such as "I am not a racist" even when these are inaccurate perceptions.
  3. Some analysts believe that there is reason for cautious optimism; throughout U.S. history subordinate racial and ethnic groups have struggled to gain the freedom and rights which were previously withheld from them movements comprised of both whites and people of color will continue to oppose racism in everyday life, to aim at healing divisions among racial groups, and to teach children about racial tolerance.

UNDERSTANDING AND ANALYZING THE BOXES

After reading the chapter and studying the outline, re-read the boxes and write down key points and possible questions for class discussion.

Sociology and Everyday Life: How Much Do You Know About Race, Ethnicity, and Sports?

Key Points:
Discussion Questions:

1.

2.

3.

Framing Race in the Media: Do Multicultural Scenes in Television Ads Reflect Reality?

Key Points:

Discussion Questions:

1.

2.

3.

You Can Make a Difference: Working For Racial Harmony

Key Points:

Discussion Questions:

1.

2.

3.

PRACTICE TESTS

MULTIPLE CHOICE QUESTIONS

Select the response that best answers the question or completes the statement:

1. Today, sociologists emphasize that race is: ____. (p. 254)
   a. a consequence of genetic inheritance.
   b. a socially constructed reality, not a biological one.
   c. a form of prejudice begun by 19th century Sociologists.
   d. a fundamentally important distinction that needs to be maintained.

2. A(n) __________ is a category of people who have been singled out as inferior or superior, often on the basis of real or alleged physical characteristics such as skin color, hair texture, eye shape, or other subjectively selected attributes. (p. 255)
   a. ethnic group
   b. status group
   c. cohort
   d. race

3. All of the following are main characteristics of ethnic groups, *except*: (p. 255)
   a. unique cultural traits.
   b. a feeling of ethnocentrism.
   c. territoriality.
   d. skin color and eye shape.

4. To classify people as Italian American, Jewish American, or Irish American is to classify them according to their __________. (p. 255)
   a. ethnic group
   b. status group
   c. cohort
   d. race

5. According to the text, the terms "majority group" and "minority group" are: (p. 258)
   a. accurate because "majority groups" always are larger in number than "minority groups."
   b. misleading because people who share ascribed racial or ethnic characteristics automatically constitute a group.
   c. less accurate terms than "dominant group" and "subordinate group," which more accurately reflect the importance of power in the relationships.
   d. no longer used because they are "politically incorrect."

6. Which of the following groups is considered to be the dominant group in North America? (p.258)
   a. WASPs
   b. DINKs
   c. Yuppies
   d. women

7. _____ is a negative attitude based on faulty generalizations about members of selected racial and ethnic groups. (p. 258)
   a. Prejudice
   b. Discrimination
   c. Stereotyping
   d. Genocide

8. Prejudice __________. (p.258)
   a. is always negative
   b. is always positive
   c. can be positive or negative
   d. is neither positive nor negative

9. Overgeneralizations about the appearance, behavior, or other characteristics of members of particular categories are called: (p. 258)
   a. discriminations.
   b. scapegoats.
   c. stereotypes.
   d. ethnic groups.

10. According to the text, the use of Native American names, images, and mascots by sports teams is an example of: (p. 259)
    a. prejudice.
    b. discrimination.
    c. stereotyping.
    d. genocide.

11. All of the following are statements of subtle racism, except: (p. 259)
    a. African American athletes which suggest that they have "natural" abilities.
    b. African Americans are better suited for team positions requiring speed and agility.
    c. African Americans lack the necessities to handle coaching and management positions.
    d. Whites are described as people who have better decision-making skills.

12. Members of subordinate racial and ethnic groups are often blamed for societal problems, such as unemployment or economic recession, over which they have no control. This is an example of __________. (p.260)
    a. the authoritarian personality
    b. institutional discrimination
    c. small-group discrimination
    d. scapegoating

13. The ultimate form of discrimination is __________. (p. 260)
    a. genocide
    b. fratricide
    c. homicide
    d. assault

14. According to sociologist Robert Merton's typology, an umpire who is prejudiced against African Americans and deliberately makes official calls against them when they are at bat, is an example of a(n): (p. 260)
    a. unprejudiced nondiscriminator.
    b. unprejudiced discriminator.
    c. prejudiced nondiscriminator.
    d. prejudiced discriminator.

15. Institutional discrimination consists of: (p. 261)
    a. one-on-one acts by members of the dominant group that harm members of the subordinate group or their property.
    b. day-to-day practices of organizations and institutions that have a harmful impact on members of subordinate groups.
    c. the division of the economy into two areas of employment, a primary sector or upper tier, and a secondary sector or lower tier.
    d. the deliberate, systematic killing of an entire people or nation.

16. Special education classes that originally were intended to provide extra educational opportunities for children with various types of disabilities but now amount to a form of racial segregation in many school districts are an example of _____ discrimination. (p. 262)
    a. indirect institutionalized
    b. direct institutionalized
    c. small group
    d. isolate

17. According to the contact hypothesis, contact between people from divergent groups should lead to favorable attitudes and behaviors when certain factors are present. Members of each group must have all of the following, *except*: (p. 262)
    a. equal status.
    b. pursuit of the same goals.
    c. being competitive with one another to achieve their goals.
    d. positive feedback from interactions with each other.

18. _____ is a process by which members of subordinate racial and ethnic groups become absorbed into the dominant culture. (p. 263)
    a. Assimilation
    b. Ethnic pluralism
    c. Accommodation
    d. Internal colonialism

19. __________ assimilation, or acculturation, occurs when members of an ethnic group adopt dominant group traits, such as language, dress, values, religion, and food preferences. (p. 263)
    a. Psychological
    b. Biological
    c. Structural
    d. Cultural

20. _____ is the coexistence of a variety of distinct racial and ethnic groups within one society. (p. 263)
    a. Assimilation
    b. Ethnic pluralism
    c. Accommodation
    d. Internal colonialism
    e. Segregation

21. Based on early theories of race relations by W. E. B. Du Bois, sociologist Oliver C. Cox suggested that African Americans were enslaved because: (p. 266)
    a. of prejudice based on skin color.
    b. they were the cheapest and best workers that owners could find for heavy labor in the mines and on plantations.
    c. they allowed themselves to be dominated by wealthy people.
    d. they agreed to work for passage to the United States, believing that they would be released after a period of indentured servitude.

22. Sociologist _____ has suggested that race, class, and cultural factors must be considered in explaining black-white relations in the United States. (p. 266)
    a. William Julius Wilson
    b. Joe R. Feagin
    c. Oliver C. Cox
    d. Robert Merton
    e. Emile Durkheim

23. All of the following theories are conflict perspectives on racial-ethnic relations, *except*: (pp. 266-267)
    a. the caste perspective.
    b. split-labor market theory.
    c. internal colonialism.
    d. ethnic pluralism.

24. According to _____ , sports reflects the interests of the wealthy and powerful; athletes are exploited in order for coaches, managers, and owners to gain high levels of profit and prestige. (pp. 266-267)
    a. symbolic interactionists
    b. sports sociologists
    c. functionalists
    d. conflict theorists
    e. postmodern theorists

25. The division of the economy into a primary sector, composed of higher paid workers in more secure jobs, and a secondary sector, composed of lower paid workers in jobs with little security and hazardous working conditions is referred to as: (p. 267)
    a. the split labor market.
    b. racial formation.
    c. gendered racism.
    d. stacking.

26. The Naturalization law of 1790 permitted only __________ immigrants to apply for naturalization (become U.S. citizens). (p.268)
    a. African-Americans
    b. white
    c. Asians
    d. Catholics

27. The experiences of Native Americans in the United States have been characterized by: (pp. 269-270)
    a. slavery and segregation.
    b. restrictive immigration laws.
    c. genocide, forced migration, and forced assimilation.
    d. internment.

28. In the winter of 1832, over half of the Cherokee Nation died during or as a result of their forced relocation from the southeastern United States to the Indian Territory in Oklahoma. This instance of forced relocation is known as the __________. (p. 270)
    a. "Dying Time"
    b. "Trail of Tears"
    c. "Native American Holocaust"
    d. "Sadness Trail"

29. Of all illegal immigrants residing in the United States, _____ are more vulnerable to deportation. (p. 279)
    a. Cuban-origin workers
    b. Puerto Rican-origin workers
    c. Dominican Republic-origin workers
    d. Mexico-origin workers

30. Since 1970, many immigrants have arrived in the United States from countries located in __________, which is the geographic region from Afghanistan to Libya and includes Arabia, Cyprus and Asiatic Turkey. (p. 280)
    a. Southeast Asia
    b. the Middle East
    c. Eastern Europe
    d. the Caucasus

TRUE-FALSE QUESTIONS

T F 1. Ethnic groups share a sense of territoriality. (p. 255)

T F 2. Race is relatively insignificant in today's society. (p. 256)

T F 3. The terms "majority group" and "minority group" are misleading because, numerically speaking, "minority" means a group that is fewer in number than a "majority" group. (p. 258)

T F 4. Use of the terms dominant and subordinate reflects the importance of power in relationships. (p.258)

T F 5. The extermination of 6 million European Jews by Nazi Germany is an example of genocide. (pp. 260-261)

T F 6. Discrimination is an individual activity; it is never based on norms of an organization or community. (p. 260)

T F 7. Equalitarian pluralism is an ideal typology that has never actually been identified in any society. (p. 263)

T F 8. The class perspective views racial and ethnic inequality as a permanent feature of U. S. society. (p. 266)

T F 9. The internal colonialism model helps explain the continued exploitation of immigrant groups such as the Chinese, Filipinos, and Vietnamese. (p. 267)

T F 10. The term "gendered racism" refers to the interactive effect of racism and sexism in the exploitation of women of color. (p. 267)

T F 11. White Anglo-Saxon Protestants (WASPS) have been the most privileged group in the United States. (p. 272)

T F 12. The term "white ethnics" was coined to identify immigrants who came from European countries other than England, such as Ireland, Poland, Germany, and Italy. (p. 275)

T F 13. White ethnics have not experienced discrimination in the United States. (p. 275)

T F 14. Most of the first Chinese, Japanese, Korean, and Filipino immigrants in the United States were men. (pp. 276-277)

T F 15. It is predicted that by 2056, the roots of the average U.S. resident will be in Africa, Asia, Hispanic countries, the Pacific Islands, and the Middle East. (p. 281)

SOCIOLOGY IN OUR TIMES: DIVERSITY ISSUES

1. Do you think Milton Bradley's statement at the beginning of Chapter 9 accurately reflects the impact of race and ethnicity on people's lives in the United States, or do you think it is an overreaction on his part? Does your answer mirror your own racial-ethnic experiences in the United States and/or other countries?

2. The next time you watch a sports event on television or in person, think about the race/ethnicity of the team members. Does your observation support (or refute) the ideas in this chapter? Where are the women, and what are they doing?

3. If we are truly honest, most of us hold at least some slight feelings of prejudice against persons we consider to be different from ourselves. But many of us do not overtly discriminate against people we have defined as being in our "out groups." Do you think it is possible to build higher levels of trust and tolerance in highly diverse societies such as the United States?

4. To what extent does the "political climate" in a country affect racial-ethnic relations? Can members of the U.S. Congress or the president bring about greater racial harmony? Can they also exacerbate racial divisiveness, strife, and social inequalities?

5. Do you considered yourself defined more strongly by your race or your ethnicity? Explain.

INTERNET EXERCISES

As you have done in previous chapters, log on to the internet, use a search engine to answer the following questions:

1. What group is the largest minority group in the United States according to the 2000 Census?

2. Explain internal colonialism, and give an example of it.

INFOTRAC COLLEGE EDITION ONLINE READINGS AND EXERCISES

Access the Infotrac website. Once you are logged on, type "black activists" in the **keyword** search field and hit return. Select the article titled "Black activists want publicist to do more to eradicate racism." Read the article and answer the following questions.

a. How did the person in the article promote racism?

b. What corrective measures did the company unveil?

Return to the search field and type "Genocide" in the search field. Next, click on any article pertaining to "genocide." Read the article and answer the following questions.

a. What is the name of the article?

b. Where is genocide happening right now and who is trying to help?

c. Would you agree with the examples given?

STUDENT CLASS PROJECTS AND ACTIVITIES

1. Investigate pluralism on your campus by conducting a brief campus study. Record your information in a written paper. Ask the following questions and include any others you may want to ask: (1) What is the number and percentage of the different ethnic or racial groups enrolled on the campus? (2) What is the number and percentage of "foreign" students? (3) What percentage of various ethnic groups hold important campus offices? Identify the office and the race or ethnicity of the office holder. (4) What percentage of racial or ethnic groups are in the campus sororities and fraternities and other organizations? (5) What percentage of racial or ethnic groups comprise both the male and female athletes who receive college or university scholarships? (6) Based on the research, evaluate your campus. How pluralistic is it? This information may be available from the admissions or registrar's office, student activities or athletic office. Write up your results following any additional instructions from your professor and submit your paper on the correct due date.

2. Conduct an informal survey of 10 people who appear to be an ethnic or racial minority in our country. Ask these specific questions and record the responses: (1) Ask the respondent's name, general address, and specific race or ethnicity; (2) Do you ever experience racism or bigotry or discrimination in some form in our society? (3) If so, what was it? (4) Can you remember the first time you were hurt by bigotry or discrimination in some way? (5) Can you remember a more recent time? (6) How did you respond? (7) How would you like to have responded? Provide any other questions that you would like to ask and then provide a summary, conclusion, and evaluation of this project. Submit your papers at the appropriate time following any other specific instructions given by your instructor.

3. Examine your own ethnicity or racial heritage, if known. If your ethnicity or race is mixed, then select the one that you believe to be the most dominant heritage in your life. If your ethnicity or race is unclear, or not known, then examine your own socialization into an ethnic or racial group. Note that many people, because of adoption or other factors, may not know their original ethnic heritage, but may be able to identify their own racial heritage. Prepare a minimum five-page report on your experience as a member of your particular group. Gather as much background information as possible in exploring your own ethnic or racial heritage. If known, trace the ethnic history of your own family, such as the approximate date of your ancestors' migration to the United States, the obstacles they faced, their cultural practices and experiences from childhood, usual activities, and family experiences. Include activities with extended kin members, type of education received, if any, and church and neighborhood experiences. Record any other event or experience that influenced your identity and membership in the group. Include any other information you desire, but do provide a conclusion and evaluation of this project. Submit your paper at the appropriate time following any other instructions provided by your professor.

ANSWERS TO PRACTICE TESTS FOR CHAPTER 9

Answers to Multiple Choice Questions

1. b Today, Sociologists emphasize that race is a socially constructed reality. (p. 254)
2. d Race is always now discussed in terms of physiology. (p. 255)
3. d All of the following are main characteristics of ethnic groups, *except:* skin color and eye shape. (p. 255)
4. a These are examples. (p. 255)
5. c According to the text, the terms "majority group" and "minority group" are less accurate terms than "dominant group" and "subordinate group," which more accurately reflect the importance of power in the relationships. (p. 258)
6. a White Anglo Saxon Protestants. (p. 258)
7. a Prejudice is a negative attitude based on faulty generalizations about members of selected racial and ethnic groups. (p. 258)
8. c Believing a group of people is better at something naturally. (p. 258)
9. c Stereotypes are over generalizations about the appearance, behavior, or other characteristics of particular categories. (p. 258)
10. c According to the text, the use of Native American names, images, and mascots by sports teams is an example of stereotyping. (p. 259)
11. c All of the following are statements of subtle racism, except the statement that African Americans lack the necessities to handle coaching and management positions; this is an example of overt racism. (p. 259)
12. d This is typically done by the people in power. (p.260)
13. a Destroying a people and their culture. (p. 260)
14. d According to sociologist Robert Merton's typology, an umpire who is prejudiced against African Americans and deliberately makes official calls against them when they are at bat, is an example of a prejudiced discriminator. (p. 260)
15. b Institutional discrimination consists of day-to-day practices of organizations and institutions that have a harmful impact on members of subordinate groups. (p. 261)
16. a Special education classes that originally were intended to provide extra educational opportunities for children with various types of disabilities, but now amount to a form of racial segregation are an indirect example of institutionalized discrimination. (p. 262)
17. c According to the contact hypothesis, contact between people from divergent groups should lead to favorable attitudes and behaviors when certain factors are present. Members of each group must have all of the following, *except* being competitive with one another to achieve their goals. (p. 262)
18. a Assimilation is a process by which members of subordinate racial and ethnic groups become absorbed into the dominant culture. (p. 263)

19 d And in the process losing their ethnic traits. (p. 263)

20. b Ethnic pluralism is the coexistence of a variety of distance racial and ethnic groups within one society. (p. 263)

21. b Based on early theories of race relations by W.E.B. Du Bois, sociologist Oliver C. Cox suggested that African Americans were enslaved because they were the cheapest and best workers that owners could find for heavy labor in the mines and on plantations. (p. 266)
22. a Sociologist William Julius Wilson has suggested that race, class, and culture factors should be considered in explaining black-white relations in the United States. (p. 266)
23. d All of the following theories are conflict perspectives on racial-ethnic relations, *except:* ethnic pluralism. (pp. 266-267)
24. d According to conflict theorists, sports reflects the interests of the wealthy and powerful; athletes are exploited in order for coaches, managers, and owners to gain high levels of profit and prestige. (pp. 266-267)
25. a The division of the economy into a primary sector, composed of higher-paid workers in more secure jobs, and a secondary sector, composed of lower-paid workers in jobs with little security and hazardous working conditions is referred to as the split-labor market. (p. 267)
26. b And who was "white" at the time was different from today. Those from eastern and southern Europe often were considered non-white. (p. 268)
27. c The experiences of Native Americans in the United States have been characterized by genocide, forced migration, and forced assimilation. (pp. 269-270)
28. b A devastating tale, where people were removed at gun point without having even packing. (p. 270)
29. d Of all illegal immigrants residing in the United States, Mexico-origin workers are more vulnerable to deportation. (p. 279)
30. b These immigrants are from the Middle East. (p. 280)

Answers to True-False Questions

1. True (p. 255)

2. False According to several analysts, race permeates every part of American life. (p. 256)

3. True (p. 258)

4. True (p. 258)

5. True (pp. 260-261)

6. False Although discriminatory actions are engaged in by individuals, some organizations and communities do prescribe actions that intentionally have a differential and negative impact on members of subordinate groups. (p. 260)

7. False Some analysts suggest that Switzerland is an example of equalitarian pluralism because over 6 million people with French, German, and Italian cultural heritages peacefully coexist there. (p. 263)

8. False The caste perspective views racial and ethnic inequality as a permanent feature of U.S. society. (p. 266)

9. False The internal colonialism model does not help explain the continued exploitation of immigrant groups such as the Chinese, Filipinos, and Vietnamese. It helps to explain the experiences of indigenous populations (such as Native Americans and Mexican Americans) who were colonized by Euro-Americans and others who invaded their lands and conquered them. (p. 267)

10. True (p. 267)

11. True (p. 272)

12. True (p. 275)

13. False White ethnics have experienced discrimination in the United States. See text for examples. (p. 275)

14. True (pp. 276-277)

15. True (p. 281)

CHAPTER 10
SEX AND GENDER

BRIEF CHAPTER OUTLINE

Sex: The Biological Dimension
- Hermaphrodites/Transsexuals
- Sexual Orientation

Gender: The Cultural Dimension
- The Social Significance of Gender
- Sexism

Gender Stratification in Historical and Contemporary Perspective
- Hunting and Gathering Societies
- Horticultural and Pastoral Societies
- Agrarian Societies
- Industrial Societies
- Postindustrial Societies

Gender and Socialization
- Parents and Gender Socialization
- Peers and Gender Socialization
- Teachers, Schools, and Gender Socialization
- Sports and Gender Socialization
- Mass Media and Gender Socialization
- Adult Gender Socialization

Contemporary Gender Inequality
- Gendered Division of Paid Work
- Pay Equity (Comparable Worth)
- Paid Work and Family Work

Perspectives on Gender Stratification
- Functionalist and Neoclassical Economic Perspectives
- Conflict Perspectives
- Feminist Perspectives

Gender Issues in the Future

CHAPTER SUMMARY

It is important to distinguish between **sex**—refers to the biological and anatomical differences between females and males—and **gender**, which refers to the culturally and socially constructed differences between females and males found in the meanings, beliefs, and practices associated with "femininity" and "masculinity." Gender is socially significant because it leads to differential treatment of men and women; sexism (like racism) often is used to justify discriminatory treatment. **Sexism** is linked to patriarchy, a hierarchical system in which cultural, political, and economic structures are male dominated. In most hunting and gathering societies, fairly equitable relationships exist because neither sex has the ability to provide all of the food necessary for survival. In horticultural societies, hoe cultivation is compatible with child care, and a fair degree of gender equality exists because neither sex controls the food supply while in pastoral societies, women have relatively low status. In agrarian societies, male dominance is very apparent; tasks require more labor and physical strength, and women are seen as too weak or too tied to child-rearing activities to perform these activities. In industrialized societies, a gap exists between unpaid work performed by women at home and paid work performed by men and women. In postindustrial societies, technology supports a service- and information-based economy; most adult women are in the labor force and have double responsibilities – those from the labor force and those from the family. The key agents of gender socialization are parents, peers, teachers and schools, sports, and the mass media, all of which tend to reinforce stereotypes of appropriate gender behavior. Gender inequality results from the economic, political, and educational discrimination of women. In most workplaces, jobs are either gender segregated or the majority of employees are of the same gender. Gender segregated occupations lead to a disparity, or **pay gap**, between women's and men's earnings. Even when women are employed in the same job as men, on average they do not receive the same pay. Many women have a "second shift" because of their dual responsibilities for paid and unpaid work. According to functional analysts, husbands perform instrumental tasks of economic support and decision making, and wives assume expressive tasks of providing affection and emotional support for the family. Conflict analysts suggest that the gendered division of labor within families and the workplace results from male control and dominance over women and resources. Although feminist perspectives vary in their analyses of women's subordination, they all advocate social change to eradicate gender inequality.

LEARNING OBJECTIVES

After reading Chapter 10, you should be able to:

1. Distinguish between sex and gender and explain their sociological significance.

2. Explain why sex is not always clear-cut; differentiate between hermaphrodites, transsexuals, and transvestites.

3. Describe the relationship between gender roles, gender identity, and body consciousness.

4. Define sexism and explain how it is related to discrimination and patriarchy.

5. Trace gender stratification from early hunting and gathering societies until to contemporary industrial societies.

6. Explain how gender relationships in both the workplace and in families change in postindustrial societies.

7. Describe the process of gender socialization and identify specific ways in which parents, peers, teachers, sports, and mass media contribute to the process.

8. Describe gender bias and explain how schools operate as a gendered institution.

9. Discuss the gendered division of paid work and explain its relationship to the issue of pay equity or comparable worth.

10. Trace changes in labor force participation by women and note how these changes have contributed to the "second shift."

11. Describe functionalist and neoclassical economic perspectives on gender stratification and contrast them with conflict perspectives.

12. State the feminist perspective on gender equality and outline the key assumptions of liberal, radical, socialist, and black (African American) feminism.

KEY TERMS (defined at page number shown and in glossary)

body consciousness 292
comparable worth 306
feminism 311
gender 292
gender bias 301
gender identity 292
gender role 292
hermaphrodite 289
homophobia 292
matriarchy 294
patriarchy 294
pay gap 305
primary sex characteristics 288
secondary sex characteristics 288
sex 288
sexism 294
sexual orientation 290
transsexual 289
transvestite 289

CHAPTER OUTLINE

I. SEX: THE BIOLOGICAL DIMENSION
    A. **Sex** refers to the biological and anatomical differences between females and males.
        1. **Primary sex characteristics** are the genitalia used in the reproductive process; **secondary sex characteristics** are the physical traits (other than reproductive organs) that identify an individual's sex.
    B. Hermaphrodites/Transsexuals are examples of when sex is not always clear-cut.
        1. **Hermaphrodites**, people whose sexual differentiation are ambiguous or incomplete, tend to have some combination of male and female genitalia.
        2. **Transsexuals** often feel they are the opposite sex from their sex organs.
        3. **Transvestites** are males who live as women or females who live as men, but do not alter their genitalia.
    C. **Sexual orientation** is a preference for emotional-sexual relationships with members of the opposite sex (heterosexuality), the same sex (homosexuality), or both (bisexuality).

II. GENDER: THE CULTURAL DIMENSION

A. **Gender** refers to the culturally and socially constructed differences between females and males found in the meanings, beliefs, and practices associated with "femininity" and "masculinity."

1. A micro level analysis of gender focuses on how individuals learn **gender roles** -- the attitudes, behavior, and activities that are socially defined as appropriate for each sex and are learned through the socialization process -- and **gender identity** -- a person's perception of the self as female or male. **Body consciousness**, how a person feels about his or her body, is a part of gender identify
2. A macrolevel analysis of gender examines structural features, external to the individual, which perpetuate gender inequality, including gendered institutions that are reinforced by a gendered belief system, based on ideas regarding masculine and feminine attributes that are held to be valid in a society.

B. The Social Significance of Gender

1. Gender is a social construction with important consequences in everyday life.
2. Gender stereotypes hold that men and women are inherently different in attitudes, behavior, and aspirations.

C. Sexism

1. **Sexism** -- the subordination of one sex, usually female, based on the assumed superiority of the other sex -- is interwoven with **patriarchy** -- a hierarchical system of social organization in which cultural, political, and economic structures are controlled by men.
2. **Matriarchy** is a hierarchical system of social organization in which cultural, political, and economic structures are controlled by women; few societies have been organized in this manner.

III. GENDER STRATIFICATION IN HISTORICAL AND COMTEMPORARY PERSPECTIVE

A. The earliest known division of labor between women and men is in *hunting* and *gathering societies*; men hunt while women gather for food. A relatively equitable relationship exists because both sexes are necessary for survival.

B. In *horticultural societies*, women make an important contribution to food production because hoe cultivation is compatible with child care; a fairly high degree of gender equality exists because neither sex controls the food supply; however in *pastoral societies*, herding primarily is done by men; women contribute relatively little to subsistence production and thus have relatively low status. Social practices, such as polygyny, contribute to gender inequality.

C. Gender inequality increases in *agrarian societies* as men become more involved in food production and as private ownership of property increases. Women are secluded and subordinated through practices of purdah and genital mutilation.

D. In *industrial societies* -- those in which factory or mechanized production has replaced agriculture as the major form of economic activity -- the status of women tends to decline further; gendered division of labor increases the economic and political subordination of women.

E. In *postindustrial societies,* technology supports a service– and information–based economy; most adult women are in the work force and have double responsibilities – those from the labor force and those from the family; more households are headed by women with no male present.

V. GENDER AND SOCIALIZATION

A. Parents and Gender Socialization
   1. From birth, parents act toward children on the basis of gender labels; children's clothing and toys reflect their parents' gender expectations.
   2. Boys are encouraged to engage in gender-appropriate behavior; they are not to show an interest in "girls'" activities.

B. Peers and Gender Socialization
   1. Peers help children learn prevailing gender-role stereotypes, as well as gender-appropriate and inappropriate behavior.
   2. During adolescence, peers often are stronger and more effective agents of gender socialization than are adults.
   3. Among college students, peers play an important role in career choices and the establishment of long term, intimate relationships.

C. Teachers and Schools and Gender Socialization
   1. From kindergarten through college, schools operate as gendered institutions; teachers provide important messages about gender through both the formal content of classroom assignments and informal interactions with students.
   2. Teachers may unintentionally demonstrate **gender bias** -- the showing of favoritism toward one gender over the other -- toward male students.

D. Sports and Gender Socialization
   1. The type of game played differs with the child's sex: from elementary school through high school, boys play football and other competitive sports while girls are cheerleaders, members of the drill team, and homecoming queens.
   2. For many males, sports participation and spectatorship is a training ground for masculinity; for females, sports still is tied to the male gender role, thus making it very difficult for girls and women to receive the full benefits of participating in such activities.

E. Mass Media and Gender Socialization
   1. Gender stereotyping is found in media, ranging from children's cartoons to adult shows.
   2. On television, more male than female roles are shown, and male characters typically are more aggressive, constructive, and direct, while females are deferential toward others or use manipulation to get their way.

3. Advertising also plays an important role in gender socialization.

F. Adult Gender Socialization

1. Men and women are taught gender-appropriate conduct in schools and the workplace.
2. Different gender socialization may occur as people reach their forties and enter "middle age."

V. CONTEMPORARY GENDER INEQUALITY

A. Gendered Division of Paid Work

1. *Gender-segregated work* refers to the concentration of women and men in different occupations, jobs, and places of work.
2. *Labor market segmentation* -- the division of jobs into categories with distinct working conditions -- results in women having separate and unequal jobs in the secondary sector of the split- or dual-labor market that are lower paying, less prestigious, and have fewer opportunities for advancement.

B. Pay Equity (Comparable Worth)

1. Occupational segregation contributes to a **pay gap** -- the disparity between women's and men's earnings.
2. **Comparable worth** is the belief that wages ought to reflect the worth of a job, not the gender or race of the worker.

C. Paid Work and Family Work

3. Although both men and women profess that working couples should share household responsibilities, researchers find that family demands remain mostly women's responsibility, even among women who hold full-time paid employment.
4. Many women have a "double day" or "second shift" because of their dual responsibilities for paid and unpaid work.

VI. PERSPECTIVES ON GENDER STRATIFICATION

A. Functional and neoclassical economic perspectives on the family view the division of family labor as ensuring that important societal tasks will be fulfilled.

1. According to the *human capital model,* individuals vary widely in the amount of education and job training they bring to the labor market.
2. From this perspective, what individuals earn is the result of their own choices and labor market demand for certain kinds of workers at specific points in time. This perspective is rooted in the premise that individuals are evaluated based on their human capital in an open, competitive market where education, training, and other job-enhancing characteristics are taken into account.

B. According to the conflict perspective, the gendered division of labor within families and the workplace results from male control of and dominance over women and resources.

1. Although men's ability to use physical power to control women diminishes in industrial societies, they still remain the head of household, control the property, and hold more power through their predominance in the most highly paid and prestigious occupations and the highest elected offices.
2. Conflict theorists in the Marxist tradition assert that gender stratification results from private ownership of the means of production; some men not only gain control over property and the distribution of goods but also gain power over women. Marriage serves to enforce male dominance. Men of the capitalist class instituted monogamous marriage in order to ensure the paternity of their offspring for inheritance of wealth and property.

C. Feminist Perspectives

1. **Feminism** refers to a belief that women and men are equal and that they should be valued equally and have equal rights.
2. In **liberal feminism**, gender equality is equated with equality of opportunity.
3. According to **radical feminists**, male domination causes all forms of human oppression, including racism and classism.
4. **Socialist feminists** suggest that women's oppression results from their dual roles as paid and unpaid workers in a capitalist economy. In the workplace, women are exploited by capitalism; at home, they are exploited by patriarchy.
5. **Multicultural feminism** is based on the belief that women of color experience a different world than other people because of multilayered oppression based on race/ethnicity, gender, and class.

VII. GENDER ISSUES IN THE FUTURE

A. In the past century, women have made significant progress in the labor force; laws have been passed to prohibit sexual discrimination in the workplace and school; women became more visible in education, the government, and the professional world.

B. Many men have joined feminist movements not only to raise their consciousness about men's concerns, realizing that what harms women may also be harmful to men, but also about the need to eliminate sexism and gender bias.

C. The **pay gap** between men and women should continue to shrink, but this may be due in part to decreasing wages paid to men or to the rise in some sages because of the economic boom for some in recent decades.

D. The burden of the "second shift", or the "double day" will probably pressure women's inequality at home and in the workplace for another generation.

ANALYZING AND UNDERSTANDING THE BOXES
After reading the chapter and studying the outline, re-read the boxes and write down key points and possible questions for class discussion.

Sociology and Everyday Life: How Much Do You Know About Body Image and Gender?

Key Points:

Discussion Questions:

1.

2.

3.

Framing Gender in the Media: "You Can Never Be too Beautiful" and Teen Plastic Surgery
Key Points:

Discussion Questions:

1.

2.

3.

You Can Make a Difference: Joining Organizations to Overcome Sexism and Gender Inequality

Key Points:

Discussion Questions:

1.

2.

3.

PRACTICE TESTS

MULTIPLE CHOICE QUESTIONS

Select the response that best answers the question or completes the statement:

1. In responding to the question, "Why do women and men feel differently about their bodies?" the text points out that: (p.288)
   a. cultural differences in appearance norms may explain women's greater concern.
   b. in the United States, the image of female beauty as childlike and thin is continually flaunted by the advertising industry.
   c. women of all racial-ethnic groups, classes, and sexual orientations regard their weight as a crucial index of their acceptability to others.
   d. all of the above.

2. Sociologists use the term _____ to refer to the biological attributes of men and women; _____ is used to refer to the distinctive qualities of men and men that are culturally created. (p. 288)
   a. gender/sex
   b. sex/gender
   c. primary sex characteristics/secondary sex characteristics
   d. secondary sex characteristics/primary sex characteristics

3. Which of the following is an example of *primary sex characteristics*? (p. 288)
   a. pubic hair
   b. enlarged breasts
   c. wider hips
   d. genitalia

4. A(n) _____ is a person in whom sexual differentiation is ambiguous or incomplete. (p. 289)
   a. hermaphrodite
   b. transsexual
   c. transvestite
   d. homosexual

5. A person's perception of the self as female or male is referred to as one's: (p. 292)
   a. self concept.
   b. gender identity.
   c. gender role.
   d. gender belief system.

6. Extreme prejudice directed at gays, lesbians, bisexuals, and others who are perceived as not being heterosexual is known as __________. (p.292)
   a. discrimination
   b. hate crime
   c. homophobia
   d. sexual phobia

7. The eating disorder known as __________ occurs when a person binges by consuming large quantities of food and then purges the food by induced vomiting, excessive exercise, laxatives, or fasting. (p. 293)
   a. obesity
   b. anorexia
   c. bulimia
   d. food phobia

8. According to historian Susan Bordo, the anorexic body and the muscled body: (p. 294)
   a. illustrates that eating problems and bodybuilding are unrelated.
   b. are opposites.
   c. are united against ideas of soft, flabby flesh.
   d. is not gendered experiences.

9. All of the following statements are correct regarding sexism, *except*: (p. 294)
   a. sexism, like racism, is used to justify discriminatory treatment.
   b. evidence of sexism is found in the under valuation of women's work.
   c. women may experience discrimination in leisure activities.
   d. men cannot be the victims of sexist assumptions.

10. __________ is a hierarchical system of social organization in which cultural, political, and economic structures are controlled by women. (p. 294)
    a. Gender organization
    b. Pastoralism
    c. Patriarchy
    d. Matriarchy

11. A society where men are seen as "natural" heads of households, presidential candidates, corporate executives, college presidents, etc., and where women are men's subordinates is an example of a __________. (p. 294)
    a. Sexual society
    b. Pastoral society
    c. Patriarchy
    d. Matriarchy

12. In most hunting and gathering societies: (p. 294)
    a. a relatively equitable relationship exists because neither sex has the ability to provide all of the food necessary for survival.
    b. women are not full economic partners with men.
    c. relationships between women and men tend to be patriarchal in nature.
    d. menstrual taboos place women in subordinate positions by monthly segregation into menstrual huts.

13. Gender inequality and male dominance became institutionalized in _____ societies. (p. 295)
    a. hunting and gathering
    b. horticultural and pastoral
    c. agrarian
    d. industrial

14. All of the following statements regarding gender socialization are correct, *except*: (p. 298)
    a. virtually all gender roles have changed dramatically in recent years.
    b. many parents prefer boys to girls because of stereotypical ideas about the relative importance of males and females to the future of the family and society.
    c. parents tend to treat baby boys more roughly than baby girls.
    d. parents' choices of toys for their children are not likely to change in the near future.

15. In their investigation of college women, anthropologists Dorothy C. Holland and Margaret A. Eisenhart determined that: (p. 300)
    a. peer groups on college campuses are not organized around gender relations.
    b. the peer system propelled women into a world of romance in which their attractiveness to men counted most.
    c. the women's peers immediately influenced their choices of majors and careers.
    d. peer pressure did not involve appearance norms.

16. Teachers who devote more time, effort, and attention to boys than to girls in schools are an example of: (pp. 300-301)
    a. gender socialization.
    b. gender identity.
    c. gender role differentiation.
    d. gender bias.

17. A comprehensive study of gender bias in schools suggested that girls' self-esteem is undermined in school through all of the following experiences, *except*: (p. 301)
    a. a relative lack of attention from teachers.
    b. sexual harassment from male peers.
    c. an assumption that girls are better in visual-spatial ability, as compared with verbal ability.
    d. the stereotyping and invisibility of females in textbooks.

18. According to the text's discussion of gender and athletics: (pp. 301-302)
    a. women who engage in activities that are assumed to be "masculine" (such as bodybuilding) may either ignore their critics or attempt to redefine the activity or its result as "feminine" or womanly.
    b. some women believe they are more likely to win women's bodybuilding competitions if they "overbuild" their bodies.
    c. being active in sports such as gymnastics makes women less likely to be the victims of anorexia and bulimia.
    d. women bodybuilders have learned that they are less likely to win competitions if they look like a fashion model.

19. According to feminist scholars, women experience __________ as a result of economic, political, and educational discrimination. (p. 303)
    a. sexual orientation
    b. gender inequality
    c. empowerment
    d. pay equity

20. __________ refers to the concentration of women and men in different occupations, jobs, and places of work. In 2000, for example, 98 percent of all secretaries in the United States were women. (p. 303)
    a. Gender bias
    b. Sexual orientation
    c. Heterogeneity
    d. Gender-segregated work

21. According to recent research by sociologist Elizabeth Higginbotham, African American professional women: (pp. 303-304)
    a. find themselves limited in employment in certain sectors of the labor market.
    b. are concentrated in public sector employment.
    c. frequently are employed as public school teachers, welfare workers, librarians, public defenders, and faculty members at public colleges.
    d. all of the above.

22. The __________ is calculated by dividing women's earning by men's to yield a percentage, also known as the earnings ratio. (p. 305)
    a. salary increment
    b. salary potential
    c. pay gap
    d. pay increment

23. The belief that wages ought to reflect the worth of a job, not the gender or race of the worker, is referred to as: (p. 306)
    a. the earnings ratio.
    b. pay equity.
    c. non-comparable worth.
    d. the pay gap.

24. According to the functionalist/human capital model: (pp. 309-310)
    a. what women earn is the result of their own choices and the needs of the labor market.
    b. the gendered division of labor in the workplace results from male control of and dominance over women and resources.
    c. women's oppression results from their dual roles as paid and unpaid workers in a capitalist economy.
    d. none of the above.

25. Critics of functionalist and neoclassical economic perspectives have pointed out that these perspectives: (p. 310)
    a. exaggerate the problems inherent in traditional gender roles.
    b. fail to critically assess the structure of society that makes educational and occupational opportunities more available to some than others.
    c. overemphasize factors external to individuals that contribute to the oppression of white women and people of color.
    d. focus on differences between men and women without taking into account the commonalities they share.

26. According to the __________ theoretical perspective, the gendered division of labor results from males having more economic, physical, political, and interpersonal power than women. (p. 310)
    a. conflict
    b. human capital
    c. neoclassical economic
    d. functionalist

27. __________ is the belief that women and men are equal and should be valued equally and have equal rights. (p. 311)
    a. The Bill of Rights
    b. Comparable worth
    c. Feminism
    d. Sexism

28. According to the __________ theoretical perspective, male domination causes all forms of human oppression, including racism and classism. (p. 311)
    a. conflict
    b. liberal feminist
    c. radical feminist
    d. functionalist

29. A women's group fights for better childcare options, women's right to choose an abortion, and the elimination of sex discrimination in the workplace, arguing that the roots of women's oppression lie in women's lack of equal civil rights and educational opportunities. The platform of this group reflects: (p. 311)
    a. black (African American) feminism.
    b. socialist feminism.
    c. radical feminism.
    d. liberal feminism.

30. Overall, women earn approximately _____ cents on the dollar compared with men. (p. 314)
    a. 37
    b. 47
    c. 67
    d. 77

TRUE-FALSE QUESTIONS

T F 1. Most "sex differences" actually are socially constructed "gender differences." (p. 292)

T F 2. The terms Homosexual and gay most often refer to males who prefer same-sex relations. (p. 290)

T F 3. Gendered belief systems generally do not change over time. (p. 293)

T F 4. Sexism is the subordination of one sex, usually female, based on the assumed superiority of the other sex. (p. 294)

T F 5. The earliest known division of labor between women and men is in agrarian societies. (p. 294)

T F 6. As societies industrialize, the status of women tends to rise. (p. 296)

T F 7. According to your textbook, most people have an accurate perception of their physical appearance. (p. 291)

T F 8. Unlike parents in other countries, parents in the United States do not prefer boys to girls. (p. 298)

T F 9. Most studies of gender socialization by parents have been based on the experiences of white, middleclass families. (p. 299)

T F 10. A person's gender is culturally created. (p. 292)

T F 11. Female peer groups place more pressure on girls to do "feminine" things than male peer groups place on boys to do "masculine" things. (p. 300)

T F 12. By young adulthood, men and women no longer receive gender-related messages from peers. (p. 300)

T F 13. Gender-appropriate behavior is learned through the socialization process. (p. 298)

T F 14. Many teachers use sex segregation as a way to organize students, resulting in unnecessary competition between females and males. (p. 301)

T F 15. The mass media are a powerful source of gender socialization. (p. 302)

SOCIOLOGY IN OUR TIMES: DIVERSITY ISSUES

1. Think about your own body consciousness and how a person perceives and feels about his or her body. What do you consider to be your most positive attributes? The most negative? How are your feelings about your body linked to the gender belief system in society?

2. According to sociologist Becky W. Thompson, eating problems exist among women of color, working-class women, lesbians, and some men. Why do many people assume that the primary victims of eating problems are white, middle-class, heterosexual women? Do you think men of color as well as white men are affected by "ideal" images of masculinity in the United States?

3. Do you consider your sex/gender to be an asset, a liability, or neither? Can racism and sexism be used to justify discriminatory treatment of some people in a society?

4. As a woman or a man, what do you think your life would have been like if you had lived in a hunting and gathering society? A horticultural or pastoral society? An agrarian society? What is the relationship between gender and daily life, public and private, in industrial and postindustrial societies?

5. Why is pay equity, or comparable worth, an important issue for all workers, not just women? (p. 306) Are you planning to enter a male-dominated job? A female-dominated job? If you are a man, are you likely to seek employment in female-dominated fields? If you are a woman, are you likely to seek employment in male-dominated fields?

INTERNET EXCERSISES

Use any of the search engines that have been provided in the previous chapters to assist you in the following:

1. Find an article that discusses the pay gap between a female and male professor at a four-year university. Notice if the pay gap exists in certain fields of study (i.e. Sociology, Business).

2. Patriarchy, as defined, is a hierarchical system in which cultural, political, and economic structures are male dominated. Clearly, individuals in certain societies practice this idea; however, there are societies that practice a matriarchy structure. Use the internet to aid you in finding an example of a matriarchal society.

INFOTRAC COLLEGE EDITION ONLINE READINGS AND EXERCISES

Access the Infotrac website in order to answer the questions in this section. Once you are logged on type in " Women's Studies and Men's Studies: Friends or Foes?" into the **keyword** search field and hit return. Click on the selection. Read the article and respond to the following questions:

a. Why is a men's studies program critical to understanding gender issues?

b. Why would women's studies scholars and activists be wary of something called "men's studies?"

Now, return to the search page and type in "Gender Issues in Advertising Language." Click on this article. Read the article and respond to the following questions:

a. According to the article, how are women portrayed in advertising? How are men described?

b. Which gender is most commonly used when advertising children's toys?

c. Give an example of a gender neutral statement that can be used to advertise a product.

STUDENT CLASS PROJECTS AND ACTIVITIES

1. Conduct an informal survey of ten students on your campus about their opinions of the women's movement and feminism. Some suggested guideline questions to ask: (1) What is your name, age, gender, and college major? (2) Do you support women's rights? Why?

Why not? (3) What is your opinion of the women's movement in our country? (4) Do you consider yourself a feminist? Why? Why not? (5) Has the women's movement improved your life in any way? If so, how? (6) Is there still a need for a strong women's movement? Why? Why not? (7) Does the present women's movement reflect the view of most women? Explain your answer. (8) Is a woman's place at home with her children, if it is economically feasible? (9) Do women today have more freedom than their mothers did? (10) Do women today enjoy life more than their mothers did? (11) Include any other information you may want to obtain from the respondents. In writing up this project, provide the responses to each particular question and a summary of the responses. Provide a conclusion and a personal evaluation of this project and submit it to your instructor at the appropriate time.

2. Research the history of sexual harassment in the United States. You are to research: (1) how the term came to be defined; (2) how it became an issue in our society; (3) how it has affected women in the workplace; (4) some specific examples of how it has affected women in the workplace; (5) the prevalence of sexual harassment in the workplace; (6) some court cases involving charges of sexual harassment and their outcomes. Include any other information in this research project. Provide in your paper a bibliography, a summary of your findings, and a personal evaluation of the project. Submit your paper to your instructor at the appropriate time.

3. This is a project that works best for a coed campus but it could easily be altered to fit any campus. You are to reverse your roles regarding dating on your college campus. The women are to initiate the date invitation either in person or by telephone, invite the young men out, at their expense. Each woman is to drive to her date's residence and pick him up. She should also help him get in and out of the car, open all doors for him, assist with sitting and with removing and replacing coats, accept the check and pay for the same. She is, in the meanwhile, to maintain a conversation about normal interests, preferably with reversed themes and roles, such as sports and the stereotyped "male talk" on campus. The female should initiate all conversation, remain in control of the conversation, interrupt when necessary to maintain control, remain aloof, when necessary, take up more space when sitting, and give all outward appearance of being completely in charge. The woman should walk her date to the door and should initiate any signs of affection, if desired. The assignment for the male is this: The males in the study should initiate any activity in order to get asked out by the females, either by females in the class or on campus. The males are to remain rather passive, after making the first move. They are to act dependent and end every statement with a question, such as "What do you think?" "Is this right?" "Is this okay with you?", etc. They are to walk in front of their dates, and ask for assistance with the doors, with the car, and with their coats or hats. They are not to pay for anything, and are to sit quietly and listen politely to their date's conversation. They can never interrupt or take charge of the conversation in any way whatsoever. The females determine all the activity. Following your date, record your specific activities, actions, and reactions and respond, in writing, to the following questions: (1) Were you comfortable? Explain. (2) Did the behaviors seem amusing? (3) What communication problems arose? (4) Did habitual patterns of behavior surface? (5) How did you feel about this activity? (6) What

specifically did you learn? (7) Record any other information, especially an evaluation of the project, and submit your paper to your instructor at the appropriate time.

ANSWERS TO PRACTICE TESTS FOR CHAPTER 10

Answers to Multiple Choice Questions

1. d In responding to the question, "Why do women and men feel differently about their bodies?" the text points out that cultural differences in appearance norms may explain women's greater concern; in the United States, the image of female beauty as childlike and think is continually flaunted by the advertising industry; and women of all racial-ethnic groups, classes, and sexual orientations regard their weight as a crucial index of their acceptability to others. Thus, "all of the above" is the correct answer. (p. 288)
2. b Sociologists use the term sex to refer to the biological attributes of men and women; gender is used to refer to the distinctive qualities of men and women that are culturally created. (p. 288)
3. d Determined for sexual identification. (p. 288)
4. a A hermaphrodite is a person in whom sexual differentiation is ambiguous or incomplete. (p.289)
5. b A person's perception of the self as female or male is referred to as one's gender identity. (p. 292)
6. c By definition. (p. 292)
7. c Not as easy to identify, but a terrible disorder. (p. 293)
8. c According to historian Susan Bordo, the anorexic body and the muscled body are united against the idea of soft, flabby flesh. (p. 294)
9. d All of the following statements are correct regarding sexism, *except*: men cannot be the victims of sexist assumptions. (p. 294)
10. d Matri, referring to females or mothers. (p. 294)
11. c Patri, referring to males or fathers. (p. 294)
12. a In most hunting and gathering societies, a relatively equitable relationship exists because neither sex has the ability to provide all of the food necessary for survival. (p. 294)
13. c Gender inequality and male dominance became institutionalized in agrarian societies. (p. 295)
14. a All of the following statements regarding gender socialization are correct, *except:* virtually all gender roles have changed dramatically in recent years. (p. 298)
15. b In their investigation of college women, anthropologists Dorothy C. Holland and Margaret A. Eisenhart determined that the peer system propelled women into a world of romance in which their attractiveness to men counted most. (p. 300)
16. d Teachers who devote more time, effort and attention to boys than to girls in schools are an example of gender bias. (pp. 300-301)
17. c A comprehensive study of gender bias in schools suggested that girls' self-esteem is undermined in school through all of the following experiences, *except:* an

assumption that girls are better in visual-spatial ability, as compared with verbal ability. (p. 301)

18. a According to the text's discussion of gender and athletics, women who engage in activities that are assumed to be "masculine" (such as bodybuilding) may either ignore their critics or attempt to redefine the activity or its result as "feminine" or womanly. (p. 306)
19. b At all levels of the hierarchy. (p. 303)
20. d Often referred to as pink collar jobs. (p. 303)
21. d According to recent research by sociologist Elizabeth Higginbotham, African American professional women find themselves limited in employment in certain sectors of the labor market; are concentrated in public sector employment; and are frequently employed as public school teachers, welfare workers, librarians, public defenders, and faculty members at public colleges. Thus, "all of the above" is the correct answer. (pp. 309-310)
22. c Women make ¾ of what men make. (p. 305)
23. b The belief that wages ought to reflect the worth of a job, not the gender or race of the worker, is referred to as pay equity. (p. 306)
24. a According to the functionalist/human capital model, what women earn is the result of their own choices and the needs of the labor market. (pp. 309-310)
25. b Critics of functionalist and neoclassical economic perspectives have pointed out that these perspectives fail to critically assess the structure of society that makes educational and occupational opportunities more available to some than others. (p. 310)
26. a typical conflict view focusing on oppression. (p. 310)
27. c Most Americans are then Feminists because most Americans believe this. (p. 311)
28. c Philosophical belief of radical feminists. (p. 311)
29. d A women's group fights for better childcare option, women's right to choose an abortion, and the elimination of sex discrimination in the workplace, arguing that the roots of women's oppression lie in women's lack of equal civil rights and educational opportunities. The platform of this group reflects liberal feminism. (p. 311)
30. d Fact. (p. 314)

Answers to True-False Questions

1. True (p. 292)

2. True (p. 290)

3. False Gendered belief systems do change over time as gender roles change. Sometimes these changes are relatively slow, however, because popular stereotypes and existing cultural norms serve to reinforce gendered institutions in society. (p. 293)

4. True (p. 294)

5. False The earliest known division of labor between women and men is in hunting and gathering societies. (p. 294)

6. False As societies industrialize, the status of women tends to decline. (p. 296)

7. False Absolutely false. (p. 291)

8. False Even in the United States, parents tend to prefer boys to girls. This is especially true for a first or only child. (p. 298)

9. True (p. 299)

10. True (p. 292)

11. False Male peer groups place more pressure on boys to do "masculine" things than female peer groups place on girls to do "feminine" things. (p. 300)

12. False As young adults, men and women still receive many gender-related messages from peers. (p. 300)

13. True (p. 298)

14. True (p. 301)

15. False If they embrace traditional notions of masculinity and femininity, their personal and social success is assured. (p. 302)

CHAPTER 11
FAMILIES AND INTIMATE RELATIONSHIPS

BRIEF CHAPTER OUTLINE

Families in Global Perspective
- Family Structure and Characteristics
- Marriage Patterns
- Patterns of Descent and Inheritance
- Power and Authority in Families
- Residential Patterns

Theoretical Perspectives on Families
- Functionalist Perspectives
- Conflict and Feminist Perspectives
- Symbolic Interactionist Perspectives
- Post Modern Perspectives

Developing Intimate Relationships and Establishing Families
- Love and Intimacy
- Cohabitation and Domestic Partnerships
- Marriage
- Housework and Child-Care Responsibilities

Child-Related Family Issues and Parenting
- Deciding to Have Children
- Adoption
- Teenage Pregnancies
- Single-Parent Households
- Two-Parent Households
- Remaining Single

Transitions and Problems in Families
- Family Transitions Based on Age and the Life Course
- Family Violence
- Children in Foster Care
- Elder Abuse
- Divorce
- Remarriage

Family Issues in the Future

CHAPTER SUMMARY

Families are relationships in which people live together with commitment, form an economic unit and care for any young, and consider their identity to be significantly attached to the group. While the **family of orientation** is the family into which a person is born and in which early socialization usually takes place, the **family of procreation** is the family a person forms by having or adopting children. Sociologists investigate marriage patterns (such as **monogamy** and **polygamy**), descent and inheritance patterns (such as **patrilineal**, **matrilineal**, and **bilateral**

descent), familial power and authority (such as **patriarchal**, **matriarchal**, and **egalitarian families**), residential patterns (such as **patrilocal**, **matrilocal**, and **neolocal residence**, and in-group or out-group marriage patterns (i.e. **endogamy** and **exogamy**). Functionalists emphasize that families fulfill important societal functions, including sexual regulation, socialization of children, economic and psychological support, and the provision of social status. By contrast, conflict and feminist perspectives view the family as a source of social inequality and focus primarily on the problems inherent in relationships of dominance and subordination. Symbolic interactionists focus on family communication patterns and subjective meanings that members assign to everyday events. Postmodern perspectives view the family as *permeable*-- being subject to modification or change. The ideal culture of the United States emphasizes *romantic love*, but women and men may not share the same perceptions about romantic love today. Families and intimate relationships have changed dramatically in the United States, with significant increases in **cohabitation**, **domestic partnerships**, and **dual-earner marriages**. Child-related family issues and parenting -- deciding to have children, adoption, teenage pregnancies, single-parent households, two-parent households, and remaining single -- began to change in the late 1950s. Families go through many transitions and experiences based on age and the life course. Stages in the life course in industrialized nations typically include infancy and childhood, adolescence and young adulthood, middle adulthood and late adulthood. But family problems, including those leading to divorce, are often too complex for solutions through legislation. Most people who divorce remarry, sometimes creating stepfamilies or **blended families**. Regardless of problems facing families today, many people still demonstrate their faith in the future by choosing to create families and to have children.

LEARNING OBJECTIVES

After reading Chapter 11, you should be able to:

1. Explain why it has become increasingly difficult to develop a concise definition of family.

2. Describe kinship ties and distinguish between families of orientation and families of procreation.

3. Compare and contrast extended and nuclear families.

4. Describe the different forms of marriage found across cultures.

5. Discuss the system of descent and inheritance, and explain why such systems are important in societies.

6. Distinguish between patriarchal, matriarchal, and egalitarian families.

7. Explain the differences in residential patterns and note why most people practice endogamy.

8. Describe functionalist, conflict and feminist, and symbolic interactionist perspectives on families.

9. Explain the postmodernist perspectives on family.

10. Describe how U.S. families have changed over the past two decades.

11. Describe cohabitation and domestic partnerships and note key social and legal issues associated with each.

12. Describe the major problems faced in dual-earner marriages, and note why the double shift most often is a problem for women.

13. Discuss the major issues associated with adoption, teenage pregnancies, single parent households, and two parent households.

14. Explain the main causes of family violence and elder abuse.

15. Discuss the major factors that contribute to the large numbers of children in foster care today.

16. Describe the major causes and consequences of divorce and remarriage in the United States.

17. Define a blended family and provide two examples.

KEY TERMS (defined at page number shown and in glossary)

bilateral descent 323
blended family 342
cohabitation 330
domestic partnerships 330
dual-earner marriages 324
egalitarian families 325
endogamy 325
exogamy 325
extended family 321
families 318
family of orientation 320
family of procreation 320
homogamy 330
kinship 319
marriage 321
matriarchal family 324
matrilineal descent 322
matrilocal residence 325
monogamy 321
neolocal residence 325
nuclear family 321
patriarchal family 323
patrilineal descent 322
patrilocal residence 324
polyandry 322
polygamy 321
polygyny 321
second shift 331
sociology of family 325

CHAPTER OUTLINE

I. FAMILIES IN GLOBAL PERSPECTIVE
   A. **Families** are relationships in which people live together with commitment, form an economic unit and care for any young, and consider their identity to be significantly attached to the group.
   B. Family Structure and Characteristics
      1. In preindustrial societies, the primary social organization is through **kinship**-a social network of people based on common ancestry, marriage, or adoption.
      2. In industrialized societies, other social institutions fulfill some functions previously taken care of by kinship ties; families are responsible primarily for regulating sexual activity, socializing children, and providing affection and companionship for family members.
      3. Many of us will be members of two types of families: a **family of orientation**, the family into which we are born or adopted and in which early socialization usually takes place, and a **family of procreation**, the family we form by having or adopting children.
      4. Extended and nuclear families
         a. An **extended family** is a family unit composed of relatives (such as grandparents, uncles, and aunts) in addition to parents and children who live in the same household.
         b. A **nuclear family** is a family composed of one or two parents and their dependent children, all of whom live apart from other relatives.
   C. Marriage Patterns
      1. **Marriage** is a legally recognized and/or socially approved arrangement between two or more individuals that carries certain rights and obligations and usually involves sexual activity.
      2. In the United States, **monogamy,** a marriage between two partners, usually a woman and a man, is the only form of marriage sanctioned by law.
      3. **Polygamy** is the concurrent marriage of a person of one sex with two or more members of the opposite sex.
         a. The most prevalent form of this marriage pattern is **polygyny,** the concurrent marriage of one man with two or more women.
         b. **Polyandry**, the marriage of one woman with two or more men, is very rare.
   D. Patterns of Descent and Inheritance
      1. In reindustrialize societies, the most common pattern of unilineal descent is **patrilineal descent** - tracing descent through the father's side of the family whereby a legitimate son inherits his father's property and sometimes his position upon the father's death.
      2. **Matrilineal descent** traces descent through the mother's side of the family; however, inheritance of property and position usually is traced from the maternal uncle (mother's brother) to his nephew (mother's son).

3. In industrial societies such as the United States, kinship usually is traced through both parents; **bilateral descent** is a system of tracing descent through both the mother's and father's sides of the family.

E. Patterns of Power and Authority in Families:
1. **Patriarchal family**: a family structure in which authority is held by the eldest male (usually the father), who acts as head of household and holds power over the women and children.
2. **Matriarchal family**: a family structure in which authority is held by the eldest female (usually the mother), who acts as head of household.
3. **Egalitarian family**: a family structure in which both partners share power and authority equally.

F. Residential Patterns:
1. **Patrilocal residence**: the custom of a married couple living in the same household (or community) as the husband's family.
2. **Matrilocal residence**: the custom of a married couple living in the same household (or community) as the wife's parents.
3. In industrialized nations, most couples hope to live in a **neolocal residence**: the custom of a married couple living in their own residence apart from both the husband's and the wife's parents.

G. Endogamy and Exogamy.
1. **Endogamy** refers to cultural norms prescribing that people marry within their own social group or category.
2. **Exogamy** refers to cultural norms prescribing that people marry outside their own social group or category.

II. THEORETICAL PERSPECTIVES ON FAMILIES

A. The *sociology of family* is the subdiscipline of sociology that attempts to describe and explain patterns of family life and variations in family structure.

B. Functionalist Perspective
1. The family is important in maintaining the stability of society and the well-being of individuals.
2. According to Emile Durkheim, both marriage and society involve a mental and moral fusion of individuals; division of labor contributes to greater efficiency in all areas of life.
3. Four key functions of families in advanced industrial societies are:
   a. sexual regulation;
   b. socialization;
   c. economic and psychological support for members; and
   d. provision of social status and reputation.

C. Conflict and Feminist Perspectives
1. Families are a primary source of inequality.
2. According to some conflict theorists, families in capitalist economies are similar to workers in a factory: women are dominated at home by men the same way workers are dominated by capitalists in factories; reproduction of children and

care for family members at home reinforce the subordination of women through unpaid (and devalued) labor.

3. Some feminist perspectives focus on patriarchy rather than class because men's domination over women existed long before private ownership of property; contemporary subordination is rooted in men's control over women's labor power.

D. Symbolic Interactionist Perspectives

1. Symbolic interactionists examine the roles of husbands, wives, and children as they act out their own parts and react to the actions of others.
2. According to Peter Berger and Hansfried Kellner, interaction between marital partners contributes to a shared reality: newlyweds bring separate identities to a marriage but gradually construct a shared reality as a couple.
3. According to Jessie Bernard, women and men experience marriage differently: there is "his" marriage and "her" marriage.

E. Postmodernist Perspectives

1. In the information age, the postmodern family is *permeable*--capable of being diffused or invaded in such a way that an entity's original purpose is modified or changed.
    a. The nuclear family is only one of many family forms.
    b. Maternal love is transformed into shared parenting.
    c. The individual values autonomy of the individual more than the family unit.
2. Urbanity is another characteristic of the postmodern family.
    a. The boundaries between the workplace and the home become more open and flexible.
    b. New communications technologies integrate and control labor.

III. DEVELOPING INTIMATE RELATIONSHIPS AND ESTABLISHING FAMILIES

A. Love and Intimacy

1. Although the ideal culture emphasizes romantic love, men and women may not share the same perceptions about love: women tend to express their feelings verbally, while men tend to express their love through nonverbal actions.
2. Love and intimacy are closely intertwined; intimacy may by psychic, sexual, or both.
3. Perceptions about sexual activity vary from one time period to another.

B. Cohabitation and Domestic Partnerships

1. **Cohabitation** refers to a couple who live together without being legally married.
2. Characteristics of persons most likely to cohabit are as follows: under age 45, have been married before, or are older individuals who do not want to lose financial benefits (such as retirement benefits) that are contingent upon not marrying.
3. Many lesbian and gay couples cohabit because they cannot enter into a legally recognized marriage; some have sought recognition of **domestic partnerships** -

household partnerships in which an unmarried couple lives together in a committed sexually intimate relationship and is granted the same benefits as those accorded to married heterosexual couples.

C. Marriage
   1. Couples marry for reasons such as being "in love," desiring companionship and sex, wanting to have children, feeling social pressure, attempting to escape from their parents' home, or believing they will have greater resources if they get married.
   2. Most people engage in **homogamy** - the pattern of individuals marrying those who have similar characteristics, such as race/ethnicity, religious background, age, education, or social class; however, some researchers claim that people look for partners whose personality traits differ from but complement their own.

D. Housework and Child-Care Responsibilities
   1. Over 50% of all U.S. marriages are **dual-earner marriages** in which both spouses are in the labor force. Over half of all employed women hold fulltime, year-round jobs.
   2. Many married women also have a **second shift**--the domestic work that employed women perform at home after they complete their workday on the job.
   3. Women employed fulltime who are single parents probably have the greatest burden of all; they have complete responsibility for the children and household, often with little or no help from ex-husbands or relatives.

IV. CHILD-RELATED FAMILY ISSUES AND PARENTING

A. Deciding to have children
   1. Couples deciding not to have children may consider themselves "child free," while those who do not produce children through no choice of their own may consider themselves "childless."
      a. Advances in birth control techniques now make it possible to determine choice of parenthood, number of children, and the spacing of their births.
      b. Some couples experience involuntary infertility; a leading cause is sexually transmitted diseases.

B. *Adoption* is a legal process through which rights and duties of parenting are transferred from a child's biological and/or legal parents to new legal parents.

C. Teenage pregnancies have decreased over the past three decades; however, they may be viewed as a crisis because an increase has occurred in the number of births among unmarried teenagers.

D. Recently, single- or one-parent households have increased significantly due to divorce and to births outside marriage.

E. Two-parent households
   1. Parenthood in the United States is idealized, especially for women.
   2. Children in two-parent families are not guaranteed a happy childhood simply because both parents reside in the same household.

F. Remaining single

1. Some never-married people remain single by choice; other never-married individuals remain single out of necessity.
2. The proportion of singles varies significantly by racial and ethnic group.

V. TRANSITIONS IN FAMILIES AND PROBLEMS IN FAMILIES

A. Family transitions based on age and the life course
   1. People are assigned different roles and positions in society based on their age and the family structure in a particular society.
   2. Stages in the contemporary life course in industrialized nations typically include infancy and childhood, adolescence and young adulthood, middle adulthood, and late adulthood.
B. Family Violence
   1. Violence between men and women in the home is often referred to as spouse abuse or domestic violence.
   2. Women are likely to be the victims of violence perpetrated by intimate partners.
C. Children in Foster Care
   1. *Foster care* refers to institutional settings or residences where adults other than a child's own parents or biological relatives serve as caregivers.
   2. Problems in their families contribute to the large numbers of children in foster care.
D. Elder Abuse
   1. *Elder abuse* refers to physical abuse, psychological abuse, financial exploitation, and medical abuse or neglect of people aged sixty-five or older; this abuse has received increasing public attention.
   2. Ageism -- prejudice and discrimination against people on the basis of age -- may be a cause of elder abuse
E. Divorce
   1. *Divorce* is the legal process of dissolving a marriage that allows former spouses to remarry if they so choose. Most divorces are based on "irreconcilable differences" (there has been a breakdown of the marital relationship for which neither partner is specifically blamed).
   2. Recent studies show that 43 percent of first marriages end in separation or divorce within 15 years.
   3. Causes of Divorce
      a. At the macrolevel, societal factors contributing to higher rates of divorce include changes in social institutions such as religion and law.
      b. At the microlevel, characteristics that appear to contribute to divorce are:
         (1) Marriage at an early age;
         (2) A short acquaintanceship before marriage;
         (3) Disapproval of the marriage by relatives and friends;
         (4) Limited economic resources;
         (5) Having a high-school education or less;
         (6) Parents who are divorced or have unhappy marriages; and
         (7) The presence of children at the beginning of the marriage.

4. Consequences of Divorce
    a. By age 16, about one in every three white and two in every three African American children will experience divorce within their families. Some children experience more than one divorce during their childhood because one or both of their parents may remarry and subsequently divorce again.
    b. Divorce changes relationships for other relatives, especially grandparents

F. Remarriage
1. Most people who divorce get remarried: more than 40% of all marriages take place between previously married brides and/or grooms, and about half of all persons who divorce before age 35 will remarry within three years.
2. Most divorced people remarry others who have been divorced. At all ages, a greater proportion of men than women remarry and often relatively soon after divorce. Among women, those who divorce at younger ages are more likely to remarry than are those who are older; women with a college degree and without children are less likely to remarry.
3. As a result of divorce and remarriage, some people become part of stepfamilies or **blended families**, which consist of a husband and wife, children from previous marriages, and (if any) children from the new marriage.

VI. FAMILY ISSUES IN THE FUTURE

A. Some people believe that the family as we know it is doomed; others believe that a return to traditional family values will save this social institution and create greater stability in society.

B. Sociologist Lillian Rubin suggests that clinging to a traditional image of families is hypocritical in light of our society's failure to support them. Some laws have the effect of hurting children whose families do not meet the traditional model. For example, cutting down on government programs which provide food and medical care for pregnant women and infants will result in seriously ill children rather than model families.

C. Psychologist Bernice Lott suggests that people's perceptions about what constitutes a family will continue to change in the next century: the family may become those persons on whom one can depend for emotional support, who are available in crises and emergencies, or who provide continuing affections, concern, and companionship.

ANALYZING AND UNDERSTANDING THE BOXES

After reading the chapter and studying the outline, re-read the boxes and write down key points and possible questions for class discussion.

Sociology and Everyday Life: How Much Do You Know About Contemporary Trends in U.S. Family Life?

Key Points:

Discussion Questions:

1.

2.

3.

Sociology in Global Perspective: Buffering Financial Hardships: Extended Families in the Global Economy

Key Points:

Discussion Questions:

1.

2.

3.

You Can Make a Difference: Providing Hope and Help for Children

Key Points:

Discussion Questions:

1.

2.

3.

PRACTICE TEST

MULTIPLE CHOICE QUESTIONS

Select the response that best answers the question or completes the statement:

1. According to the text, traditional definitions of the family: (p. 318)
   a. are still highly applicable to today's families.
   b. include all persons in a relationship who wish to consider themselves a family.
   c. need to be expanded to provide a more encompassing perspective on what constitutes a family.
   d. are indistinguishable from contemporary definitions of the family.

2. A __________ is a relationship in which people live together with commitment, form an economic unit and care for any young, and consider their identity to be significantly attached to the group. (p. 318)
   a. cohort
   b. family
   c. secondary group
   d. culture

3. Census data show that the marriage rate has __________ in the United States since 1960. (p. 321)
   a. gone down by about two-thirds
   b. gone down by about one-third
   c. stayed about the same
   d. increased by about one-third

4. Less than _____ percent of U.S. family households are composed of a married couple with one or more children under age eighteen. (p. 320)
   a. 5
   b. 25
   c. 45
   d. 75

5. A social network of people based on common ancestry, marriage, or adoption is known as: (p. 319)
   a. kinship.
   b. a family.
   c. a clan.
   d. subculture.
   e. friendship

6. In preindustrial societies, the primary form of social organization is through __________ ties. (p. 319)
   a. occupational
   b. kinship
   c. organizational
   d. secular

7. Recent studies have shown that adult children of divorced parents are __________ to dissolve their own marriages than they were two decades ago. (p. 320)
   a. less likely
   b. equally as likely
   c. more likely
   d. (there are not enough data to determine this)

8. The family of procreation is defined as "the family _____________." (p. 320)
   a. into which a person is born
   b. in which people receive their early socialization
   c. that is composed of relatives in addition to parents and children who live in the same household
   d. a person forms by having or adopting children
   e. that lives in close proximity to other family members

9. The __________ is the family into which a person is born and in which early socialization usually takes place. (p. 320)
   a. family of procreation
   b. family of orientation
   c. kinship unit
   d. ancestry unit

10. Families that include grandparents, uncles, aunts, or other relatives who live in close proximity to the parents and children are known as a(n): (p. 321)
    a. clan.
    b. extended family.
    c. nuclear family.
    d. family of procreation.
    e. family of close proximity

11. Persons in nontraditional family structures may include those with blood ties and legal ties, but it may also include __________, which are those who are not actually related by blood but who are accepted as family members. (p. 320)
    a. optional kin
    b. fictive kin
    c. honorary members
    d. acting members

12. A(n) __________ is a family unit composed of relatives in addition to parents and children who live in the same household. (p. 321)
    a. nuclear family
    b. adoptive family
    c. generation
    d. extended family

13. _____ is the concurrent marriage of one man with two or more women, while ________ is the concurrent marriage of one woman with two or more men. (p. 321)
    a. Monogamy polygamy
    b. Patriarchy matriarchy
    c. Polygyny polyandry
    d. Polyandry polygyny

14. __________ is very rare; when it does occur, it is typically found in societies where men greatly outnumber women because of high rates of female infanticide. (p. 322)
    a. Polyandry
    b. Monogamy
    c. Serial monogamy
    d. Polygyny

15. The most prevalent pattern of power and authority in families is: (p. 324)
    a. matriarchy.
    b. monarchy.
    c. oligarchy.
    d. patriarchy.

16. All of the following statements are correct regarding power and authority in families, *except:* (pp. 323-324)
    a. Recently, there has been a trend toward more egalitarian family relationships in a number of countries.
    b. Some degree of economic independency makes it possible for women to delay marriage.
    c. Power and authority are not important issues among gay and lesbian couples because their relationships already are more egalitarian.
    d. Scholars have found no historical evidence to indicate that true matriarchies ever existed.

17. The custom of a married couple living in their own residence apart from both the husband's and the wife's parents is known as: (p. 325)
    a. isolated residence.
    b. neolocal residence.
    c. neutrallocal residence.
    d. exogamous residence.

18. When a person marries someone who comes from the same social class, racial/ethnic group, and religious affiliation, sociologists refer to this marital pattern as: (p. 325)
    a. endogamy.
    b. exogamy.
    c. inbreeding.
    d. intraclass reproduction.

19. According to _______, marriage is a microcosmic replica of the larger society. (p. 326)
    a. Emile Durkheim
    b. Jessie Bernard
    c. Max Weber
    d. Karl Pillemer

20. According to functionalists, all of the following are key functions of families, *except:* (p. 326)
    a. provision of social status.
    b. economic and psychological support.
    c. maintenance of workers so that they can function effectively in the workplace.
    d. sexual regulation and socialization of children.

21. According to _______ theorists, interaction between marital partners contributes to a shared reality. (p. 327)
    a. feminist
    b. conflict
    c. functionalist
    d. symbolic interactionist

22. Some ______analysts focus on the effect of class conflict on the family, while some _____ perspectives focus on patriarchy. (pp. 326-327)
    a. symbolic interactionist; feminist
    b. feminist; symbolic interactionists
    c. functionalist, conflict
    d. conflict; feminist

23. Sociologist ______ pointed out that women and men experience marriage differently -- as "her" marriage and "his" marriage. (p. 327)
    a. Alice Rossi
    b. Peter Berger
    c. Jessie Bernard
    d. Alfred Kinsey

24. Kissing is viewed as positive in __________ cultures. (p. 328)
    a. all of the world's
    b. primarily Western
    c. primarily African and Asian
    d. primarily North American

25. Which of the following statements regarding cohabitation is true? (p. 330)
    a. Attitudes about cohabitation have not changed very much in the past two decades.
    b. The Bureau of the Census recently developed a more inclusive definition of cohabitation.
    c. People most likely to cohabit are those who are under age 30 and who have not been married before.
    d. In the United States, many lesbian and gay couples cohabit because they cannot enter into a legally recognized marital relationship.

26. Sociologist ___________ coined the term "second shift" to refer to the domestic work that employed women perform at home after they complete their workday on the job. (p. 331)
    a. Arlie Hochschild
    b. Talcott Parsons
    c. Kath Weston
    d. Francesca Cancian

27. Recent studies of teen pregnancy have concluded that: (p. 334)
    a. teenage pregnancies are increasing rapidly in the United States.
    b. teenage mothers often are as skilled at parenting as older mothers.
    c. teen birth rates have not declined as rapidly as rates for older women.
    d. the media no longer focuses on "the problem" of teenage pregnancies.

28. ______ is generally considered to begin at age 65. (p. 337)
    a. middle adulthood
    b. late adulthood
    c. middle age
    d. late-late adulthood

29. All of the following are cited in the text as primary social characteristics of those most likely to get divorced, *except:* (p. 341)
    a. marriage at a later age and being set in one's ways.
    b. a short acquaintanceship before marriage.
    c. disapproval of marriage by relatives and friends.
    d. parents who are divorced or have unhappy marriages.

30. A _______ family consists of a husband and wife, children from previous marriages and children (if any) from the new marriage. (p. 343)
    a. single parent family
    b. nuclear family
    c. blended family
    d. traditional family

TRUE-FALSE QUESTIONS

T F 1. Most U.S. family households are composed of a married couple with one or more children under age 18. (p. 321)

T F 2. In industrialized societies, other social institutions fulfill some of the functions previously taken care of by the kinship network. (p. 319)

T F 3. By definition, "marriage" must be a legally recognized and/or socially approved arrangement. (p. 321)

T F 4. The most prevalent form of polygamy is polygyny. (pp. 321-322)

T F 5. In industrial societies, kinship is usually traced through patrilineal descent. (p. 322)

T F 6. A domestic partnership is a household partnership in which an unmarried couple lives together in a committed, sexually intimate relationship that is granted the same rights and benefits as those accorded to married heterosexual couples. (p. 330)

T F 7. About 30 percent of all marriages in the United States are dual-earner marriages. (p. 330)

T F 8. Infertility is defined as the inability to conceive after a year of unprotected relations (p. 332)

T F 9. The U.S. society has a pronatalist bias. (p. 332)

T F 10. Fewer infants are available for adoption today in the United States than in the past. (p. 333)

T F 11. Less than 1 million gay men in the United States and Canada are fathers. (p. 335)

T F 12. Maternity is the mark of adulthood for women, whether or not they are employed. (p. 335)

T F 13. African American men have a higher rate of singlehood than do Latinos and whites. (p. 336)

T F 14. Between one and two million older people in the United States are the victims of physical or mental abuse each year. (p. 339)

T F 15. Most divorced people remarry and remarry others who have been divorced. (p. 342)

SOCIOLOGY IN OUR TIMES: DIVERSITY ISSUES

1. Which of the family patterns described in this chapter are applicable to your own family? What do you consider to be the strengths and weaknesses of the family structure in which you grew up? Did changes in that structure occur over time? If so, how did those changes affect you?

2. Why do you think that many of the Baby Boom generation who declared that they did not want children have changed their minds in their thirties and forties and are now having children? What are some of the most important factors in considering to remain child-free? What are the differences in the terms "child-free" lifestyle and "childless" life-style? What values are implicated in these terms?

3. According to the text, "across race and class, numerous studies have confirmed that domestic work remains primarily women's work." Apply your sociological imagination to respond to this statement, how are "family" problems related to beliefs and values embedded in the larger society?

4. Throughout the chapter, the unique problems of lesbian and gay couples are discussed. Do you think these problems will be resolved in the future? Or, do you think these problems may become worse? Do you think our society will become more or less tolerant of diverse family patterns in the future?

INTERNET EXERCISES

Use the site of the U.S. Census to assist you in the following exercises. This site is among the more stable, but do note that web site addresses change frequently.

1. Check this site for information on U.S. households and families:
   http://www.infoplease.com/ipa/A0104688.html
   What does this data suggest?

2. Check this site to obtain resource information for American families:
   http://family.go.com
   Note: Did you surmise that this is a Disney-sponsored site?

INFOTRAC COLLEGE EDITION ONLINE READINGS AND EXERCISES

Access the Infotrac College library. Once logged on, type "Why Monogamy" in the **keyword** search field and hit return. Read the article and answer the following:

a. What are two major theories about the historical shift in many countries from polygamy to monogamy?

b. Why does monogamy increase with economic development under democracy?

Return to the **keyword** search field and enter "Polyandry." Scroll down to the article titled "On a Panhuman Preference for Monandry: Is Polyandry an Exception." Read the article and answer the following questions:

a. What is the global frequency of polyandry?

b. Wherever polyandry occurs, does polygyny co-occur? Why? Why not?

STUDENT CLASS PROJECTS AND ACTIVITIES

1. Select and then research a topic of unconventional reproduction, such as *in vitro* fertilization, (also known as test-tube fertilization), artificial insemination, or surrogacy. After selecting the topic, (1) trace the history of the development of that specific type of reproduction; (2) provide a description and definition of the reproduction procedure; (3) describe the widespread use of the procedure; and (4) discuss any complications, moral, legal, or medical that have surfaced related to the practice. Additionally, (5) examine and report any court cases that relate to your specific topic that have surfaced. In writing your paper, ensure that you provide a conclusion and a personal evaluation of this specific project.

2. Write a paper on the American family in the future. Provide at least four bibliographic sources (such as the text, other books, periodicals, newspaper articles, etc.). The paper could center on the topic: "The American Family in the New Millennium." Issues to be included are the following: (1) What type of family form will be dominant -- has the nuclear family disappeared? (2) As families change, has the look of the community

changed as well? (3) What is the status of commuter marriage? (4) Describe the dating practice: will the dating game be a "dangerous sport" because of the risk? (5) What about life-expectancy? -- has there been an extension of the life span? (6) What about "variations of the theme family" -- are there a variety of new categories, such as gay and lesbian couples with and without children, single women having babies by donor insemination, etc? (7) What about diet? (8) What type of child care is available? (9) What is the size of the family? (10) Who is taking care of aged relatives? (11) Describe the place of residence. Include any other information that pertains to this topic. In writing this paper, follow any other specific guidelines and submit your paper at the appropriate time.

3. This is a short assignment that should be typed. Be sure and submit your paper to your instructor on the due date. This paper should be a minimum of five pages. Instructions: Construct a "sociological" family tree of your relatives of at least three generations (your grandparents, parents, aunts, uncles, and your own generation) Record names, dates, and information as you gather this information. Examine your geneology for some of the following norms of marriage: (1) who married, who did not; (2) who did your relatives marry; (3) age at marriage; (4) did the norms of homogamy operate; (5) did they marry "up" or "down"; (6) number of children in the marriage; (7) number of divorces, if any; and (8) their religion. You may need to interview several family members, to get the correct information for the above. Record all information carefully. After recording the information, examine how closely your family's marital choices follow the patterns of the typical American family explained in the text and why variations might have occurred in your family tree. In your paper, include your reactions to your research on your family.

ANSWERS TO PRACTICE TEST, CHAPTER 11

Answers to Multiple Choice Questions

1. c According to the text, traditional definitions of the family need to be expanded to provide a more encompassing perspective on what constitutes a family. (p. 318)
2. b This is the ever changing definition of the family. It changes as the family changes. (p. 318)
3. b Yet, marriage is as popular as it has ever been. (p. 321)
4. b Factual numbers. (p. 320)
5. a A social network of people based on common ancestry, marriage, or adoption is known as kinship. (p. 319)
6. b This changes in industrial society. (p.319)
7. a As divorce has become normal, the stigma associated with it is having less effect. (p. 320)
8. d The family of procreation is defined as "the family a person forms by having or adopting children." (p. 320)
9. b The classic definition. (p. 320)
10. b Families that include grandparents, uncles, aunts, or other relatives who live in close proximity to the parents and children are known as an extended family. (p. 321)

11. a This relates to the constant changing found within the concept of family. (p. 320)
12. b This has been true throughout the centuries in various cultures. (p. 321)
13. c Polygyny is the concurrent marriage of one man with two or more women, while polyandry is the concurrent marriage of one woman with two or more men. (p. 321)
14. a Because of the low status of females. (p. 322)
15. d The most prevalent pattern of power and authority in families is patriarchy. (p. 324)
16. c All of the following statements are correct regarding power and authority in families, *except*: power and authority are not important issues among gay and lesbian couples because their relationships already are more egalitarian. (pp. 323-324)
17. b The custom of a married couple living in their own residence apart from both the husband's and the wife's parents is known as neolocal residence. (p. 325)
18. a When a person marries someone who comes from the same social class, racial/ethnic group, and religious affiliation, sociologists refer to this marital pattern as endogamy. (p. 325)
19. a According to Emile Durkheim, marriage is a microcosmic replica of the larger society. (p. 326)
20. c According to functionalists, all of the following are key functions of families, *except:* maintenance of workers so that they can function effectively in the workplace. (p. 326)
21. d According to symbolic interactionist theorists, interaction between marital partners contributes to a shared reality. (p. 327)
22. d Some conflict analysts focus on the effect of class conflict on the family, while some feminist perspectives focus on patriarchy. (pp. 326-327)
23. c Sociologist Jessie Bernard pointed out that men and women experience marriage differently. (p. 327)
24. b Different norms for different cultures. (p. 328)
25. d Which of the following statements regarding cohabitation is true? Answer: In the United States, many lesbian and gay couples cohabit because they cannot enter into a legally recognized marital relationship. (p.330)
26. a Sociologist Arlie Hochschild coined the term "second shift" to refer to the domestic work that employed women perform at home after they complete their workday on the job. (p. 331)
27. c Recent studies of teen pregnancy have concluded teen birth rates have not declined as rapidly as rates for older women (p. 334)
28. b Late adulthood is generally considered to begin at age 65. (p. 337)
29. a All of the following are cited in the text as primary social characteristics of those most likely to get divorced, *except:* marriage at a later age and being set in one's ways. (p. 341)
30. c The blended family is one that results from divorce and remarriage and consists of a husband and wife, children from previous marriages, and children, if any, from the new marriage. (p. 343)

Answer to True-False Questions

1. False Only 25 percent of all family households were composed of married couples with one or more children under age 18. (p. 321)

2. True (p. 319)

3. True (p. 321)

4. True (pp. 321-322)

5. False In industrial societies, kinship is usually traced through bilateral descent, a system of tracing descent through both the mother's and father's sides of the family. (p. 322)

6. True (p. 330)

7. False Over 50 percent of all marriages in the United States are dual-earner marriages. (p. 330)

8. True (p. 332)

9. True (p. 332)

10. True (p. 333)

11. False Between 1 and 3 million gay men in the United States and Canada are fathers. (p. 335)

12. True (p. 335)

13. True (p. 336)

14. True (p. 339)

15. True (p. 342)

CHAPTER 12
EDUCATION AND RELIGION

BRIEF CHAPTER OUTLINE

CHAPTER SUMMARY

Education and religion are powerful and influential forces in contemporary societies; however, there is a lack of consensus in the United States regarding the appropriate relationship between public education and religion. **Education** is the social institution responsible for the systematic transmission of knowledge, skills, and cultural values within a formally organized structure. In addition to teaching the basics, U.S. schools today also teach a myriad of topics ranging from computer skills to sex education. Functionalists have suggested that education performs a number of essential functions for society; however, conflict theorists emphasize that education perpetuates class, racial-ethnic, and gender inequalities. Symbolic interactionists point out that education may become a *self-fulfilling* prophecy for students who come to perform up or down to the expectations held for them by teachers. The U.S. public schools today are a microcosm of many of the problems facing the country: unequal funding of public schools, school violence, students dropping out, racial segregation and resegregation of schools. In analyzing the opportunities and challenges in college and universities, we find that in the U.S. today, for students who complete high school, access to colleges and universities is determined by prior academic record and the ability to pay. One of the fastest growing areas of U.S. higher education today are the community colleges, which educate about half of the nation's undergraduates and which are more affordable than the typical four-year college. By providing more flexible class schedules, part-time students have ample opportunities for attending classes. However, limited resources are one of the major problems facing community colleges today- - a problem experienced by many four-year colleges and universities as well. The four-year colleges and universities offer students a general education curriculum that exposes them to multiple disciplines and ways of knowing and offers them a variety of degrees. Problems in higher education include the soaring cost of a college degree and racial and ethnic differences in enrollment and lack of faculty diversity. **Religion** is a system of beliefs, symbols, and rituals, based on some **sacred** or supernatural realm that guides human behavior, gives meaning to life, and unites believers into a community. Although religion seeks to answer important questions such as why we exist, why people suffer and die, and what happens when we die, increases in scientific knowledge have contributed to **secularization**, the process by which religious beliefs, practices, and institutions lose their significance in sectors of society and culture. According to functionalists, religion provides meaning and purpose to life, promotes social cohesion and a sense of belonging, and provides social control and support for the government. From a conflict perspective, religion can have negative consequences in that the capitalist class uses religion as a tool of domination to mislead workers about their true interests. However, Max Weber believed that religion could be a catalyst for social change. Symbolic interactionists examine the meanings that people give to religion and the meanings they attach to religious symbols in their everyday life. Contemporary religious organizations may be categorized as **ecclesia**, **churches**, **denominations**, **sects**, and **cults**. Religion in the United States is very diverse; pluralism and religious freedom are among the important cultural values. The rise of new fundamentalism and the growth of the electronic church and the Internet has transformed religion into an international issue. The debates over curriculum, no general standards, and achievement remain controversial issues in education. Maintaining an appropriate balance between the social institutions of education and religion will be an important challenge for the United States in the future.

LEARNING OBJECTIVES

After reading Chapter 12, you should be able to:

1. Explain why education and religion are powerful and influential forces in contemporary society.

2. Describe the functionalist perspective on education and note the societal importance of manifest and latent functions fulfilled by this social institution.

3. Describe conflict perspectives on education and note how they differ from a functionalist perspective.

4. Discuss symbolic interactionist perspectives on education, and describe the significance of the self-fulfilling prophecy and labeling on educational achievement.

5. List and discuss major problems in elementary and secondary schools today.

6. Discuss the opportunities and challenges in colleges and universities in the United States.

7. List and discuss the four categories of religion.

8. Describe the functionalist perspective on religion and discuss its major functions in societies.

9. Compare and contrast civil religion with other forms of religion.

10. Describe conflict perspectives on religion and distinguish between the approaches of Karl Marx and Max Weber.

11. Describe symbolic interactionist perspectives on religion and explain how religion may be viewed differently by women and men.

12. Distinguish between the different types of religious organizations and note why a religious group may move from one category to another over time.

13. Describe recent trends in religion in the United States and explain their potential impact on public education in the future.

KEY TERMS (defined at page number shown and in glossary)

animism 366
church 372
civil religion 369
credentialism 357
cult 373
cultural capital 354
denomination 372
ecclesia 371
education 348
hidden curriculum 356
profane 365
religion 365
sacred 365
sect 372
secularization 367
tracking 354

CHAPTER OUTLINE

I. AN OVERVIEW OF EDUCATION AND RELIGION:
    A. Education and religion are powerful and influential institutions that impart values, beliefs, and knowledge considered essential to the social reproduction of individual personalities and entire cultures.

B. Education and religion are socializing institutions: early socialization primarily takes place in families and friendship networks; later socialization occurs in more formalized organizations created for the purposes of education and religion.

C. **Education** is the social institution responsible for the systematic transmission of knowledge, skills, and cultural values within a formally organized structure.

II. SOCIOLOGICAL PERSPECTIVES ON EDUCATION

A. Functionalists view education as one of the most important components of society.

1. Education serves five major *manifest* functions- open, stated, and intended goals or consequences of activities within an organization or institution:
   a. Socialization;
   b. Transmission of culture;
   c. Social control;
   d. Social placement; and
   e. Change and innovation.
2. Education has at least three *latent* functions -- hidden, unstated, and sometimes unintended consequences of activities within an organization or institution:
   a. Restricting some activities;
   b. Matchmaking and production of social networks; and
   c. Creation of a generation gap.

B. According to conflict theorists, schools perpetuate class, race, ethnic, and gender inequalities as some groups seek to maintain their privileged position at the expense of others.

1. Cultural Capital and Class Reproduction are vehicles for determining educational achievement.
   a. According to Pierre Bourdieu, children have less chance of academic success when they lack **cultural capital** -- social assets that include values, beliefs, attitudes, and competencies in language and culture.
   b. Children from middle- and upper-income families are endowed with more cultural capital than children from working-class and poverty level families.
2. Tracking and Social Inequality are closely linked
   a. Ability grouping and **tracking** - - the assignment of students to specific courses and educational programs based on their test scores, previous grades or both - - affect student's academic achievements and career choices.
   b. Tracking systems may result in students dropping out of school or their lacking the courses required to attend college.
3. The **hidden curriculum** is the transmission of cultural values and attitudes, such as conformity and obedience to authority, through implied demands found in rules, routines, and regulations of schools.
   a. Lower-class students may be disqualified from higher education and the credentials needed in a society that emphasizes **credentialism-** a process of social selection in which class advantage and social status are linked to the possession of academic qualification.

b. **Credentialism** is closely related to meritocracy: persons who acquire the appropriate credentials for a job are assumed to have gained the position through what they know.
c. Gender bias is embedded in both the formal and the hidden curricula of schools; the educational opportunities of U.S. females are not equal to those of males in their social class.

C. Symbolic Interactionist Perspective on Education
1. Labeling and the Self-Fulfilling Prophecy
a. Labeling is the process whereby a person is identified by others as possessing a specific characteristic or exhibiting a certain pattern of behavior.
b. For some students, schooling may become a *self-fulfilling prophecy*- an unsubstantiated belief or prediction that results in behavior which makes the originally false belief come true.
2. Using Labeling Theory to Examine the IQ Debate
a. IQ testing has resulted in labeling of students (e.g., African American and Mexican American children have been placed in special education classes on the basis of IQ scores when they could not understand the tests); terms such as *learning disabled* have become social constructs that lead to *stigmatization* and may be incorporated into the everyday interactions of teachers, students, and parents.
b. Labeling students based on IQ scores has been an issue for many decades. Immigrants from Eastern Europe had particularly lower IQ scores than did those arriving from northern Europe. For many of these, the IQ testing became a self-fulfilling prophecy.

III. PROBLEMS WITHIN ELEMENTARY AND SECONDARY SCHOOLS

A. Unequal funding of Public Schools
1. Most educational funds are derived from local property taxes and state legislative appropriations with the federal government paying a small amount largely for special programs.
2. Children living in affluent suburbs often attend relatively new schools and have access to the latest equipment which students in central city schools and poverty-ridden rural areas lack.
3. Recently Voucher systems have gained many supporters; others argue that this would harm rather than support public education.

B. School Violence
1. In the 1990's, violent acts resulted in numerous deaths in schools across the nation, such as in Mississippi, Kentucky, and Colorado.
2. Many students must learn in an academic environment that is similar to a maximum security prison.
3. John Devine argues that violence in schools will not end until guns in the US. are eliminated.

C. Dropping Out
 1. Dropout rates vary by race, ethnic, class and regional differences.
 2. Explanations for the group having highest rate of dropouts (Hispanics, or Latinos/as) vary:
  a. "Hispanic" or "Latino/a" incorporates a wide diversity of young people.
  b. Labeled as troublemakers, they actually become "dropouts."
  c. After being excluded from meaningful education, they have been pushed out of the educational system rather than dropping out.

D. Racial Segregation and Resegregation of Schools
 1. In many areas of the U.S., schools remain segregated or have become resegregated after earlier attempts at integration failed.
  a. Five decades after the 1954 U.S. Supreme Court ruled in *Brown v. The Board of Education of Topeka, Kansas* that segregated schools were unconstitutional, racial segregation remains.
  b. Racial segregation is increasing in many school districts and efforts to bring about desegregation or integration have failed in many school districts.
 2. Racially segregated housing patterns are associated with the high rate of school segregation.
  a. Even in supposedly integrated schools, ability grouping produces resegregation at the classroom level.
  b. Many children of color are placed in lower-level courses and special education classes.

IV. OPPORTUNITES AND CHALLENGES IN COLLEGES AND UNIVERSITIES

A. Opportunities and Challenges in Community Colleges
 1. Access to colleges and universities is determined not only by a person's prior academic record but also by the ability to pay.
 2. Community colleges are one of the fastest growing areas of U.S. higher education today.
 3. They are more affordable and offer more flexible class schedules than the typical four-year college.
 4. Limited resources are one of the major problems facing the community college today.

B. Opportunities and Challenges in Four-Year Colleges and Universities
 1. They offer a variety of degrees, ranging from the bachelor's degree to the master's to the doctorate to the professional degree.
 2. Most provide a liberal education by offering a general education curriculum.
 3. Major problems include cost of higher education, problems with students completing a degree program, racial and ethnic differences in enrollment, and lack of faculty diversity.

C. The Soaring Cost of College Education
   1. The costs of attending public institutions and four-year colleges have increased dramatically over the past decade.
   2. Enrollment of low income students in higher education has dropped since the 1980's as a result of declining scholarship funds and of many students having to hold a full- or part-time job to finance their education.
   3. Although higher education may be a source of upward mobility for talented young people from poor families, the U.S. system of higher education is significantly stratified that it may also reproduce the existing class structure.

D. Racial and Ethnic Differences in Enrollment
   1. The "Census Profiles" reflects the disproportionately few number of people of color enrolled in higher education.
   2. Minority group members accounted for only about 18 percent of local doctorates awarded in 1999-2000.

V. RELIGION IN HISTORICAL PERSPECTIVE

A. Religion and the Meaning of Life
   1. **Religion** is a system of beliefs, symbols, and rituals, based on some sacred or supernatural realm, which guides human behavior, gives meaning to life, and unites believers into a community.
   2. Religion seeks to answer important questions such as why we exist, why people suffer and die, and what happens when we die.
   3. Sacred and Profane
      a. According to Emile Durkheim, **sacred** refers to those aspects of life that are extraordinary or supernatural; those things that are set apart as "holy."
      b. Those things people do not set apart as sacred are referred to as **profane**, the everyday, secular or "worldly," aspects of life.
   4. In addition to beliefs, religion also is comprised of symbols and rituals -- symbolic actions that represent religious meanings that range from songs and prayers to offerings and sacrifices.
   5. Religions have been classified into four main categories based on their dominant belief:
      a. *Simple supernaturalism* is the belief that supernatural forces affect people's lives either positively or negatively.
      b. **Animism** is the belief that plants, animals, or other elements of the natural world are endowed with spirits or life forces having an impact on events in society.
      c. *Theism* is a belief in a god or gods.
         (1) *Monotheism* is a belief in a single, supreme being or god who is responsible for significant events such as the creation of the world. Examples include Christianity, Judaism, and Islam.
         (2) *Polytheism* is a belief in more than one god. An example is Shinto.
      d. *Transcendent idealism* is a belief in sacred principles of thought and conduct, such as truth, justice, life, and tolerance for others.

B. Religion and Scientific Explanation

1. During the Industrial Revolution, rapid growth in scientific and technological knowledge gave rise to the idea that science ultimately would answer questions that previously had been in the realm of religion.
2. Many scholars believed that scientific knowledge would result in **secularization** -- the process by which religious beliefs, practices, and institutions lose their significance in sectors of society and culture but others point out a resurgence of religious beliefs and an unprecedented development of alternative religions in recent years.

VI. SOCIOLOGICAL PERSPECTIVES ON RELIGION

A. Functionalist Perspectives on Religion

1. According to Emile Durkheim, the central feature of all religions is the presence of sacred beliefs and rituals that bind people together in a collectivity.
2. Religion has three important functions in any society:
   a. *providing meaning and purpose to life;*
   b. *promoting social cohesion and a sense of belonging;* and
   c. *providing social control and support for the government.*
3. **Civil religion** is the set of beliefs, rituals, and symbols that make sacred the values of the society and place the nation in the context of the ultimate system of meaning.

B. Conflict Perspectives on Religion

1. According to Karl Marx, the capitalist class uses religious ideology as a tool of domination to mislead the workers about their true interests; thus, religion is the "opiate of the masses."
2. By contrast, Max Weber argued that religion could be a catalyst to produce social change.
   a. In *The Protestant Ethic and the Spirit of Capitalism*, Weber linked the teachings of John Calvin with the growth of capitalism.
   b. Calvin emphasized *predestination* -- the belief that all people are divided into two groups, the saved and the damned, and only God knows who will go to heaven (the elect) and who will go to hell, even before they are born.
   c. Because people cannot know whether they will be saved, they look for signs that they are among the elect. As a result, people work hard, save their money and do not spend it on worldly frivolity; instead, they reinvest it in their land, equipment, and labor.
   d. As people worked ever harder to prove their religious piety, structural conditions became right in Europe for the industrial revolution, free markets, and the commercialization of the economy, which worked hand-in-hand with their religious teachings.

D. Symbolic interactionist Perspectives on Religion

1. For many people, religion serves as a reference group to help them define themselves. Religious symbols, for example, have a meaning to large bodies of people (e.g., the Star of David for Jews; the crescent moon and star for Muslims; and the cross for Christians).

2. Her Religion and His Religion: All people do not interpret religion in the same way. Women and men may belong to the same religions, but their individual religion will not necessarily be a carbon copy of the group's entire system of beliefs.

VII. TYPES OF RELIGIOUS ORGANIZATION

A. Some countries have an official or state religion known as an **ecclesia-** a religious organization that is so integrated into the dominant culture that it claims as its membership all members of a society. Examples include: the Anglican Church (the official Church of England), the Lutheran Church in Sweden and Denmark, the Catholic Church in Spain, and Islam in Iran and Pakistan.

B. The Church-Sect Typology.

1. A **church** is a large, bureaucratically organized religious organization that tends to seek accommodation with the larger society in order to maintain some degree of control over it.
2. Midway between the church and the sect is a **denomination:** a large, organized religion characterized by accommodation to society but frequently lacking in ability or intention to dominate society.
3. A **sect** is a relatively small religious group that has broken away from another religious organization to renew what it views as the original version of the faith.

C. A **cult** is a religious group with practices and teachings outside the dominant cultural and religious traditions of a society.

1. Some major religions (including Judaism, Islam, and Christianity) and some denominations (such as the Mormons) started as cults.
2. Cult leadership is based on charismatic characteristics of the individual, including an unusual ability to form attachments with others.

VIII. TRENDS IN RELIGION IN THE UNITED STATES

A. The rise of a new fundamentalism has occurred at the same time that a number of mainline denominations have been losing membership.

B. Some members of the political elite in Washington have vowed to bring religion "back" into schools and public life.

C. "New-right" fundamentalists are especially critical of secular humanism.

IV. EDUCATION AND RELIGION IN THE FUTURE

A. Education will remain an important social institution in the future. Also remaining, however, will be the controversies over what should be taught and how to raise levels of academic achievement in the United States.

B. Religious organizations will continue to be important in the lives of many people; however, the influence of religious beliefs and values will be felt even by those who claim no religious beliefs of their own.

C. In other nations, the rise of *religious nationalism* has led to the blending of strongly held religious and political beliefs.

D. In the United States, the influence of religion will be evident in ongoing battles over school prayer, abortion, gay rights, and women's issues, among others. On some fronts, religion may unify people; on others, it may contribute to confrontations among individuals and groups.

ANALYZING AND UNDERSTANDING THE BOXES

After reading the chapter and studying the outline, re-read the boxes and write down key points and possible questions for class discussion.

Sociology and Everyday Life: How Much Do You Know About the Impact of Religion on U.S. Education?

Key Points:

Discussion Questions:

1.

2.

3.

Framing Education in the Media: Do High School Students Rule in Media Land?

Key Points:

Discussion Questions:

1.

2.

3.

You Can Make a Difference: Reaching Out to Youth! College Student Tutors

Key Points:

Discussion Questions:

1.

2.

3.

PRACTICE TESTS

MULTIPLE CHOICE QUESTIONS

Select the response that best answers the question or completes the statement:

1. ____ is the social institution responsible for the systematic transmission of knowledge, skills, and cultural values within a formally organized structure. (p. 348)
   a. Religion
   b. Mass media
   c. The government
   d. Education

2. In the "Scopes monkey trial," Tennessee substitute high school biology teacher John Scopes was found guilty of __________. (p. 348)
   a. sexual misconduct
   b. teaching about evolution
   c. teaching about religion
   d. stealing several monkeys

3. Which of the following is a manifest function of education? (pp. 351-352)
   a. Creation of a generation gap
   b. Restricting some activities
   c. Matchmaking and production of social networks
   d. Social control

4. _____ functions are hidden, unstated, and sometimes unintended consequences of activities within an organization or institution. (p. 352)
   a. Manifest
   b. Dormant
   c. Latent
   d. Covert

5. Sociologist __________ has suggested that students come to school with differing amounts of cultural capital. (p. 353)
   a. Emile Durkheim
   b. Pierre Bourdieu
   c. Jeannie Oakes
   d. Clifford Geertz
   e. Karl Marx

6. The assignment of students to specific courses and educational programs based on their test scores, previous grades, or both is known as: (p. 354)
   a. tracking.
   b. the hidden curriculum.
   c. equitable assessment.
   d. class reproduction.
   e. mainstreaming

7. Which of the following is a criticism that has been made of *tracking* in schools? (pp. 354-356)
   a. poor and minority students receive a diluted program
   b. students' social class often influences which track they are placed in
   c. tracking can negatively affect students' career choices and academic achievement
   d. all of the above

8. According to the ______ perspective, the hidden curriculum affects working-class and poverty level students more than it does students from middle and upper-income families. (p. 356)
   a. functionalist
   b. conflict
   c. symbolic interactionist
   d. feminist

9. According to the text, schools for students from elite families: (p. 356)
   a. work to develop students' analytical powers and critical thinking skills.
   b. focus on creative activities in which students express their own ideas.
   c. have decreased in number during the past decade.
   d. have lowered their tuition in recent years in order to increase minority enrollment.

10. According to the text, the average yearly tuition for four-year colleges: (p. 363)
    a. increased more than the overall rate of inflation
    b. decreased more than the overall rate of inflation increased.
    c. remained about the same as the over rate of annual in inflation.
    d. fluctuated from year to year; ranging from little increase to great increase to a gradual decrease in the last decade.

11. According to the book The Bell Curve by Richard J. Herrnstein and Charles Murray (1994), intelligence is __________. (p. 358)
    a. the product of social factors
    b. genetically inherited
    c. higher in white Americans than in Asians
    d. higher in African Americans than in white Americans

12. According to labeling theory, labeling certain minority students as learning disabled may result in __________. (p. 358)
    a. stigmatization
    b. a self-fulfilling prophecy
    c. tracking
    d. (all of the above)

13. Most educational funds are derived from: (p. 359)
    a. the federal government.
    b. private resources.
    c. local property taxes and state legislative appropriations.
    d. students' tuition and fees.

14. According to education scholar John Devine, violence in schools will not end until __________. (p. 360)
    a. more police are assigned to school patrol duty
    b. guns are eliminated from U.S. society
    c. violence on television and in popular music is eliminated
    d. schools focus more of their curricula on moral values

15. Which of the following statements regarding higher education is true? (p. 364)
    a. The enrollment of low income students has increased since the 1980s.
    b. There has been an increase in scholarship funds over the past decade.
    c. The ability to pay for a college education reproduces the class system.
    d. African American enrollment as a percentage of total college attendance has increased over the past two decades.

16. According to Emile Durkheim, ______ refers to those aspects of life that are extraordinary or supernatural. (p. 365)
    a. religion
    b. sacred
    c. profane
    d. superhuman

17. People often act out their religious beliefs in the form of __________, which are symbolic actions that represent religious meanings. (p. 365)
    a. superstitions
    b. rituals
    c. sects
    d. cults

18. Three of the major world religions Christianity, Judaism, and Islam are characterized as: (p. 366)
    a. simple supernaturalism.
    b. animism.
    c. polytheism.
    d. monotheism.

19. Which of the following categories of religion is based on a belief in a god or gods? (p. 366)
    a. simple supernaturalism
    b. animism
    c. theism
    d. transcendent idealism

20. According to the functionalist perspective, religion offers meaning for the human experience by: (pp. 368-369)
    a. providing an explanation for events that create a profound sense of loss on both an individual and a group basis.
    b. offering people a reference group to help them define themselves.
    c. reinforcing existing social arrangements, especially the stratification system.
    d. encouraging the process of secularization.

21. Celebrations on Memorial Day and the Fourth of July are examples of: (p. 369)
    a. religious tolerance.
    b. civil religion.
    c. patriotic ethnocentrism.
    d. separation of church and state.

22. According to ________, the capitalist class uses religious ideology as a tool of domination. (p. 398)
    a. Emile Durkheim
    b. C. Wright Mills
    c. Karl Marx
    d. Max Weber

23. In regard to religion, Max Weber asserted that: (p. 370)
    a. church and state should be separated.
    b. religion could be a catalyst to produce social change.
    c. religion retards social change.
    d. the religious teachings of the Catholic church were directly related to the rise of capitalism.
    e. religion is always dysfunctional

24. In his book The Protestant Ethic and the Spirit of Capitalism, __________ asserted that the religious teachings of John Calvin were directly related to the rise of capitalism. (p. 370)
    a. Karl Marx
    b. Emile Durkheim
    c. Max Weber
    d. Daniel Patrick Moynihan

25. "Her religion" and "his religion" have been examined from a(n) ____________ perspective. (p. 371)
    a. functionalist
    b. neoMarxist
    c. conflict
    d. symbolic interactionist

26. According to feminist scholar Charlotte Perkins Gilman, religious thought and doctrine have always tended to be dominated by ______. (p. 371)
    a. bishops
    b. cardinals
    c. the middle class
    d. men

27. The Anglican church in England and the Lutheran church in Sweden are examples of a(n): (p. 372)
    a. church.
    b. sect.
    c. denomination.
    d. ecclesia.

28. In a _____, membership largely is based on birth, and children of members typically are baptized as infants. (p. 372)
    a. church
    b. sect
    c. denomination
    d. cult

29. A __________ is a relatively small religious group that has broken away from another religious organization to renew what it views as the original version of the faith. (p. 373)
    a. cult
    b. sect
    c. church
    d. ecclesia

30. According to the text, religious nationalism has led to the blending of strongly held religious and political beliefs is especially strong today in: (p. 376)
    a. the United States.
    b. Middle Eastern nations.
    c. Japan.
    d. Brazil.

TRUE-FALSE QUESTIONS

T F 1. Debates about the appropriate relationship between public education and religion have occurred only recently in the United States. (p. 348)

T F 2. According to Box 12.1, the Constitution of the United States originally specified that religion should be taught in public schools. (pp. 349-351)

T F 3. Manifest functions in education include teaching specific subjects, such as science, history, and reading. (pp. 351-352)

T F 4. Tracking and ability grouping may produce resegregation at the classroom level. (pp. 355-356)

T F 5. Meritocracy is the process of social selection in which class advantage and social status are linked to the possession of academic qualifications. (p. 357)

T F 6. Through reading materials, classroom activities, and treatment by teachers and peers, female students learn that they are less important than male students. (p. 357)

T F 7. A self-fulfilling prophecy can result from labeling, according to the research of Rosenthal and Jacobson. (p. 358)

T F 8. The issue of IQ and race/ethnicity originated with the highly controversial book by Herrnstein and Murray. (p. 358)

T F 9. Across cultures and in different eras, a wide variety of things have been considered sacred. (p. 365)

T F 10. Secularization is the process by which religious beliefs, practices, and institutions lose their significance in sectors of society and culture. (p. 367)

T F 11. According to functionalist theorists, religious teachings and practices help promote social cohesion by emphasizing shared symbolism. (p. 369)

T F 12. Karl Marx wrote *The Protestant Ethic and the Spirit of Capitalism* to explain how religion may be used by the powerful to oppress the powerless. (p. 370)

T F 13. According to Charlotte Gilman, men's religion taught people to think and realistically confront situations. (p. 371)

T F 14. Denominations tend to be more tolerant and less likely than churches to expel or excommunicate members. (p.372)

T F 15. "New-right" fundamentalists have encouraged secular humanism in the schools. (pp. 373-374)

SOCIOLOGY IN OUR TIMES: DIVERSITY ISSUES

1. According to Pierre Bourdieu, what is the relationship between cultural capital and students' educational opportunities? Do you agree or disagree? Why?

2. Has there been a "hidden curriculum" in the schools you have attended? If this hidden curriculum exists, how is it related to social class and gender bias?

3. What is meant by "her" religion and "his" religion? What are the effects of this dual system?

4. Analyze your own college or university based on the discussion of class, race, and social reproduction in higher education. Do your findings tend to confirm the assertions of conflict theorists? Why or why not?

INTERNET EXERCISES

Use the Internet to help you better understand the following concepts:

The Supreme Court recently ruled on the issue of the voucher system.

1. Who voted in support of this decision and who voted against?

2. Who dissented from this notion?

3. If given the opportunity, which way would you have decided? Explain.

INFOTRAC COLLEGE EDITION ONLINE READINGS AND EXERCISES

Access the Infotrac College Edition for the following questions. Once logged on, type "Educating children for tomorrow's world" into the **keyword** search field and hit enter. Next click on the article and read the selection and answer the questions.

a. Explain the ten major trends facing the world future society.

b. Why will these seismic shifts have great impact in the schools?

Now return to the **keyword** search field and type "Six Flags Over Israel--Orlando's Holy Land ". Next click on this article. Read the selection and answer the following questions:

a. What is "Orlando's Holy Land?"

b. What are some of the objections to "Holy Land?"

c. Why do many support "Holy Land?"

STUDENT CLASS PROJECTS AND ACTIVITIES

1. Research the topic of public education in several high-income countries, such as France, Japan, New Zealand, the Netherlands, Sweden, England, Norway, Canada, or Germany. In writing your paper, include the following information: (1) a description of their educational system; (2) a description of their specific or unique features; (3) a description of their curriculum; (4) a description of some of the strengths and weaknesses of their program; (5) a critique of their programs -- which do you think work well? Which do not? (6) any other information that will enhance our understanding of their education; (7) provide a bibliography of your references; and (8) turn your paper in to your instructor at the appropriate time.

2. Research a particular cult found in the United States or elsewhere. You should, in this research project, provide in your paper: (1) a definition of the cult; (2) the purpose of the cult; (3) a general description of the cult which should include its origin and its membership and method of acquiring members; and (3) the nature of its group life, including advantages and disadvantages of membership in the cult. After providing this basic information, include a two-page paper on a day in the life of the cult. Pretend to be a member and take the membership point of view in the writing up of a "day in the life of this cult." This day could be a very special day, such as your wedding day, or a typical school, work, or ceremonial day. Submit your paper to your instructor on the appropriate date, following any additional instructions you have been given.

3. Conduct an informal survey of any ten people in order to obtain their opinion of the ordination of women in our society. You must provide this basic information about your respondents: (1) the name, age, and address; (2) if possible, their religious affiliation (if any); (3) their gender. Provide responses to the following questions: (1) Do you support the entry of women into the seminary schools? Why? Why not? (2) Do you support the ordination of women? Why? Why not? (3) Should women be allowed to serve as full-time professional clergy or rabbi or cantor? Why? Why not? (4) What do you think should be the primary role of women in the church? (5) What do you think should be the role of women who are clergy (pastor, rabbi, cantor, priest)? (6) Do you support the decision of the Anglican Church of England to ordain women? (7) Do you support the position of the Roman Catholic Church, which is to not allow women to become priests? Why? Why not? (8) Do you have any other statements you would like to make regarding the issue of women and the church? Summarize your findings and provide a conclusion statement per each respondent. Include any raw data that you have collected. Submit your paper to your instructor at the appropriate time.

ANSWERS TO PRACTICE TESTS, CHAPTER 12

Answers to Multiple Choice Questions

1. d — Education is the social institution responsible for the systematic transmission of knowledge, skills, and cultural values within a formally organized structure. (p. 348)
2. b — This was a controversial case of which we have seen recently similar complaints made by students objecting to material being taught even on college campuses. (p. 348
3. d — Social control is a manifest function of education. (pp. 351-352)
4. c — Latent functions are hidden, unstated, and sometimes unintended consequences of activities within an organization or institution. (p. 352)
5. b — Sociologist Pierre Bourdieu has suggested that students come to school with differing amounts of cultural capital. (p. 353)
6. a — The assignment of students to specific course and educational programs based on their tests scores, previous grades, or both is known as tracking. (p. 354)

7. d All of the above apply. (pp. 354-356)
8. b According to the conflict perspective, the hidden curriculum affects working class and poverty level students more than it does students from middle and upper income families. (p. 356)
9. a. According to the text, schools for students from elite families work to develop students' analytical powers and critical thinking skills. (p. 356)
10. a Increases in average yearly tuition for four-year colleges are higher than the overall rate of inflation. (p. 363)
11. b Their analysis uses 19th Century science techniques and are not consistent with their analysis at that. (p. 358)
12. d All of the above apply. (p. 358)
13. c Most educational funds are derived from local property taxes and state legislative appropriations. (p. 359)
14. b Devine emphasizes that schools reflect the larger tumultuous society. (p. 360)
15. c Which of the following statements regarding higher education is true? The ability to pay for a college education reproduces the class system. (p. 364)
16. b According to Emile Durkheim, sacred refers to those aspects of life that are extraordinary or supernatural. (p. 365)
17. c The secular is profane. (p. 365)
18. d Three of the major world religions Christianity, Judaism, and Islam are characterized as monotheism. (p. 366)
19. b The key is the use of the word 'animism' which will clue you into the background. (p. 366)
20. a According to the functionalist perspective, religion offers meaning for the human experience by providing an explanation for events that create a profound sense of loss on both an individual and a group basis. (pp. 368-369)
21. b Celebrations on Memorial Day and the Fourth of July are examples of civil religion. (p. 369)
22. c According to Karl Marx, the capitalist class uses religious ideology as a tool of domination. (p. 369)
23. b In regard to religion, Max Weber asserted that religion could be a catalyst to produce social change. (p. 370)
24. c Weber argued that the religious teachings of Calvin were directly related to the rise of capitalism (p.370)
25. d "Her religion" and "his religion" have been examined from an symbolic interactionist perspective. (p. 371)
26. d There are a few exceptions, but typically religion has been controlled by men. (p. 371)
27. d The Anglican church in England and the Lutheran church in Sweden are examples of an ecclesia. (p. 372)
28. a In a church, membership largely is based on birth, and children of members typically are baptized as infants. (p. 372)
29. a Also remember that who is a cult is also determined by whom is doing the defining. (p. 373)

30. b According to the text, religious nationalism has led to the blending of strongly held religious and political beliefs is especially strong today in Middle Eastern nations. (p. 376)

Answers to True-False Questions

1. False Debates about the appropriate relationship between public education and religion have occurred for many years. (p. 348)

2. False Due to the diversity of religious backgrounds of the early settlers, no mention of religion was made in the original constitution. (pp. 349-351)

3. True (p. 351-352)

4. True (pp. 355-356)

5. False Credentialism is the process of social selection in which class advantage and social status are linked to the possession of academic qualifications. (p. 357)

6. True (p. 357)

7. True (p. 358)

8. False The issue of IQ and race/ethnicity first arose in regard to early twentieth-century immigrants from Southern and Eastern Europe who scored lower, on average, on IQ tests than did Northern Europeans. (p. 358)

9. True (p. 365)

10. True (p. 367)

11. True (p. 369)

12. False Max Weber wrote The Protestant Ethic and the Spirit of Capitalism to explain how religion may be a catalyst for social change. (p. 370)

13. False According to Gilman, men's religion taught people to submit and obey rather than to think and realistically confront situations. (p. 371)

14. True (p. 372)

15. False "New-right" fundamentalists have been especially critical of secular humanism in the schools. (pp. 373-374)

CHAPTER 13
POLITICS AND THE ECONOMY IN GLOBAL PERSPECTIVE

BRIEF CHAPTER OUTLINE

Politics, Power, and Authority
- Power and Authority
- Ideal Types of Authority

Political Systems in Global Perspective
- Monarchy
- Authoritarianism
- Totalitarianism
- Democracy

Perspectives on Power and Political Systems
- Functionalist Perspectives: The Pluralist Model
- Conflict Perspectives: Elite Models

The U.S. Political System
- Political Parties and Elections
- Political Participation and Voter Apathy
- Governmental Bureaucracy

Economic Systems in Global Perspective
- Preindustrial, Industrial, and Postindustrial Economies
- Capitalism
- Socialism
- Mixed Economies

Work in the Contemporary United States
- Professions
- Other Occupations
- Contingent Work
- Unemployment
- Labor Unions and Worker Activism
- Employment Opportunities for Persons with a Disability

Politics and the Economy in the Future

CHAPTER SUMMARY

Politics is the social institution through which power is acquired and exercised by some people or groups. **Government** is the formal organization that has the legal and political authority to regulate the relationships among members of a society and between the society and those outside its borders. The **state** is the political entity that possesses a legitimate monopoly over the use of force within its territory to achieve its goals. The relationship between politics and power is a strong one in all countries. **Power** -- the ability of persons or groups to carry out their will even when opposed by others -- is a social relationship involving both leaders and followers. Most leaders seek to legitimate their power through **authority** -- power that people accept as legitimate rather than coercive. According to Max Weber, there are three types of authority: (1)

charismatic, (2) **traditional**, and (3) **rational-legal (bureaucratic).** There are four main types of contemporary political systems: **monarchy**, **authoritarianism**, **totalitarianism**, and **democracy**. In a **democracy** the people hold the ruling power either directly or through elected representatives. There are two key perspectives on how power is distributed in the United States. According to the **pluralist model**, power is widely dispersed throughout many competing interest groups. According to the **elite model**, power is concentrated in a small group of elites and the masses are relatively powerless. The *power elite* is comprised of influential business leaders, key government leaders, and the military. **Political parties** are organizations whose purpose is to gain and hold legitimate control of government. The Democrats and the Republicans have dominated the U.S. political system since the mid-nineteenth century, although party loyalties have been declining in recent years. People learn political attitudes, values, and behaviors through **political socialization**. The vast governmental bureaucracy is a major source of power. The military bureaucracy is so wide ranging that it encompasses the **military industrial complex:** the mutual interdependence of the military establishment and private military contractors. Problems with the military-industrial complex are linked to problems in the economy. The **economy** is the social institution that ensures the maintenance of society through the production, distribution, and consumption of goods and services. Pre-industrial economies are characterized by **primary sector production** in which workers extract raw materials and natural resources from the environment and use them without much processing. Industrial economies engage in **secondary sector production**, which is based on the processing of raw materials (from the primary sector) into finished goods. The economic base in postindustrial economies shifts to **tertiary sector production** -- the provision of services rather than goods. As an ideal type, **capitalism** has four distinctive features: private ownership of the means of production, pursuit of personal profit, competition, and lack of government intervention. **Socialism** is characterized by public ownership of the means of production, the pursuit of collective goals, and centralized decision making. In a **mixed economy**, elements of a capitalist, market economy are combined with elements of a command, socialist economy. Mixed economies are often referred to as **democratic socialism** -- an economic and political system that combines private ownership of some of the means of production, governmental distribution of some essential goods and services, and free elections. A **Welfare state** exists where there is extensive government action to provide support and services to its citizens. For many people, jobs and professions are key sources of identity. **Occupations** are categories of jobs that involve similar activities at different work sites. **Professions** are high-status, knowledge-based occupations characterized by abstract, specialized knowledge, autonomy, self-regulation, and a degree of altruism. The **primary labor market** consists of high-paying jobs with good benefits that have some degree of security and the possibility of future advancement, while the **secondary labor market** consists of low-paying jobs with few benefits and very little job security or possibility for future advancement. **Marginal jobs** are those that do not comply with the following employment norms: job content should be legal, jobs should be covered by government regulations, jobs should be relatively permanent, and jobs should provide adequate hours and pay in order to make a living. **Contingent work** is part-time work, temporary work, and subcontracted work that offers advantages to employers but may be detrimental to workers. **Unemployment** remains a problem for many workers. Labor unions seek to improve the work environment. In 1990, the United States became the first country to address the issue of equity for persons with a disability. Many futurists predict a greater concentration of both political and economic power in the hands of a few in the new global economy. A final vision of the future is based on new challenges- such as terrorism, war, and economic instability and the way in which politics and government are conducted.

LEARNING OBJECTIVES

After reading Chapter 13, you should be able to:

1. Explain the relationship between politics, government, and the state, and note how political sociology differs from political science.

2. Distinguish between power and authority, and describe the three major types of authority.

3. Compare and contrast governments characterized by monarchy, authoritarianism, totalitarianism, and democracy.

4. State the major elements of the pluralist (functionalist) model of power and political systems.

5. State the major elements of elite (conflict) models of power and political systems, and note how they differ from pluralist (functionalist) models.

6. Describe the purpose of political parties and analyze how well U.S. parties measure up to the ideal type characteristics of political parties.

7. Explain the relationship between political socialization, political attitudes, and political participation.

8. Discuss the characteristics of the federal bureaucracy and explain what is meant by the "permanent government."

9. Describe the military industrial complex and explain why it is called an iron triangle.

10. Identify and compare and contrast pre-industrial, industrial, and postindustrial economies.

11. Describe the four distinctive features of "ideal" capitalism and explain why pure capitalism does not exist.

12. Compare and contrast capitalism, socialism, mixed economies, and a welfare state.

13. Discuss the major characteristics of professions and describe the process of deprofessionalization.

14. Distinguish between the primary and secondary labor markets and describe the types of jobs found in each.

15. Identify the occupational categories considered to be marginal jobs and explain why they are considered to be marginal.

16. Discuss contingent work and identify the role subcontracting often plays in contingent work.

17. Distinguish between the various types of unemployment and explain why the unemployment rate may not be a true reflection of unemployment in the United States.

18. Trace the development of labor unions and describe some of the means by which workers resist working conditions they consider to be oppressive.

19. Discuss employment opportunities for persons with a disability in the United States.

KEY TERMS (defined at page number shown and in glossary)

authoritarianism 386
authority 382
capitalism 396
charismatic authority 383
conglomerates 398
contingent work 405
corporations 396
democracy 386
democratic socialism 402
economy 394
elite model 389
government 380
interlocking corporate directorates 398
marginal jobs 405
military-industrial complex 394
mixed economy 402
monarchy 385
occupations 403
oligopoly 397
pluralist model 388
political action committees 389
political party 391
political socialization 392
politics 380
power 381
primary labor market 404
primary sector production 394
professions 402
rational-legal authority 383
representative democracy 387
routinization of charisma 383
secondary labor market 404
secondary sector production 395
shared monopoly 397
socialism 400
state 380
subcontracting 406
tertiary sector production 395
totalitarianism 386
transnational corporations 383
unemployment rate 406
welfare state 402

CHAPTER OUTLINE

I. POLITICS, POWER, AND AUTHORITY
   A. **Politics** is the social institution through which power is acquired and exercised by some people and groups.

B. In contemporary societies, the primary political system is the **government** -- the formal organization that has the legal and political authority to regulate the relationships among members of a society and between the society and those outside its borders.

C. Sociologists often refer to the government as the **state**--the political entity that possesses a legitimate monopoly over the use of force within its territory to achieve its goals.

D. While political science primarily focuses on power and the distribution of power in different types of political systems, **political sociology** examines the nature and consequences of power within or between societies and focuses on the social circumstances of politics and the interrelationships between politics and social structures.

E. Power and Authority
   1. **Power** is the ability of persons or groups to achieve their goals despite opposition from others.
   2. **Authority** is power that people accept as legitimate rather than coercive.
   3. According to Max Weber, there are three ideal types of authority:
      a. ***Traditional authority*** is power that is legitimized on the basis of long-standing custom.
      b. ***Charismatic authority*** is power legitimized on the basis of a leader's exceptional personal qualities or by demonstrating extraordinary insight and accomplishment that inspires loyalty and obedience from followers.
      c. ***Rational-legal authority*** is power legitimized by law or written rules and regulations.

II. POLITICAL SYSTEMS IN GLOBAL PERSPECTIVE

A. Emergence of Political Systems
   1. Political institutions first emerge in agrarian societies as they acquire surpluses and develop greater social inequality; when a city's power extended to adjacent areas, a *city-state* developed.
   2. *Nation-states* are political organizations that have recognizable national boundaries within which their citizens possess specific legal rights and obligations; approximately 190 nation-states currently exist worldwide.

B. **Monarchy** is a political system in which power resides in one person or family and is passed from generation to generation through lines of inheritance.

C. **Authoritarianism** is a political system controlled by rulers who deny popular participation in government.

D. **Totalitarianism** is a political system in which the state seeks to regulate all aspects of people's public and private lives.

E. **Democracy** is a political system in which the people hold the ruling power either directly or through elected representatives; in a **representative democracy,** citizens elect representatives to serve as bridges between themselves and the government.

III. PERSPECTIVES ON POWER AND POLITICAL SYSTEMS

A. Functionalist Perspectives: According to the **pluralist model,** power in political systems is widely dispersed throughout many competing *special interest groups* -- political coalitions comprised of individuals or groups that share a specific interest they wish to protect or advance with the help of the political system.

1. Key elements of this model:
    a. Decisions are made on behalf of the people by leaders who engage in a process of bargaining, accommodation, and compromise.
    b. Competition among leadership groups (such as leaders in business, labor, education, law, medicine, consumer groups, and government) protects people by making the abuse of power by any one group more difficult.
    c. People can influence public policy by voting in elections, participating in existing special interest groups, or forming new ones to gain access to the political system.
    d. Power is widely dispersed in society; leadership groups that wield influence on some decisions are not the same groups which may be influential in other decisions.
2. Over the last four decades, *special interest groups* have become more involved in "single-issue politics," such as abortion, gun control, gay and lesbian rights, or environmental concerns.
3. **Political action committees (PACs)** are organizations of special interest groups that fund campaigns to help elect (or defeat) candidates based on their stances on specific issues.

B. Conflict Perspectives: According to the **elite model**, power in political systems is concentrated in the hands of a small group of elites and the masses are relatively powerless.

1. Key elements of this model:
    a. Decisions are made by the elite, possessing greater wealth, education, status, and other resources than do the "masses" it governs.
    b. Consensus exists among the elite on the basic values and goals of society; however, consensus does not exist among most people in society on these important social concerns.
    c. Power is highly concentrated at the top of a pyramid shaped social hierarchy; those at the top of the power structure come together to set policy for everyone.
    d. Public policy reflects the values and preferences of the elite, not the preferences of the people.
2. According to C. Wright Mills, the power elite is comprised of leaders at the top of business, the executive branch of the federal government, and the military (especially the "top brass" at the Pentagon). The elites have similar class backgrounds and interests.
    a. The corporate rich are the most powerful because of their unique ability to parlay the vast economic resources at their disposal into political power.
    b. At the middle level of the pyramid, Mills placed the legislative branch of government, interest groups, and local opinion leaders.

c. The bottom (and widest layer) of the pyramid is occupied by the unorganized masses who are relatively powerless and vulnerable to economic and political exploitation.

3. G. William Domhoff referred to elites as the *ruling class* -- a relatively fixed group of privileged people who wield sufficient power to constrain political processes and serve underlying capitalist interests.
   a. The ruling class constitutes less than one percent of the U.S. population.
   b. The corporate rich and their families influence the political process in three ways:
      (1) they influence the candidate selection process by helping to finance campaigns and providing favors to political candidates;
      (2) through participation in the special interest process, they are able to gain favors, tax-breaks, regulatory rulings, and other governmental supports;
      (3) they gain access to the policy-making process by holding prestigious positions on governmental advisory committees, presidential commissions, and other governmental appointments.

IV. THE U.S. POLITICAL SYSTEM

A. A **political party** is an organization whose purpose is to gain and hold legitimate control of government.
   1. Parties develop and articulate policy positions; educate voters about the issues and simplify the choices for them; recruit candidates who agree with those policies, and help those candidates win office; and, when elected, hold the candidates responsible for implementing the party's policy positions.
   2. Since the Civil War, two political parties -- the Democratic and the Republican -- have dominated the political system in the United States.
      a. Although both parties have been successful in getting their candidates elected at various times, they generally do not meet the ideal characteristics for a political party because they do not offer voters clear policy alternatives.
      b. Moreover, the two parties are oligarchies, dominated by active elites who hold views that are further from the center of the political spectrum than are those of a majority of members of their primary.

B. Political Participation and Voter Apathy
   1. **Political socialization** is the process by which people learn political attitudes, values, and behavior. For young children, the family is the primary agent of political socialization.
   2. Socioeconomic status affects people's political attitudes, values, beliefs, and participation.
   3. The United States has one of the lowest percentages of voter turnout of all Western nations; slightly over 51 percent of the voting-age population voted in the 2000 presidential election.

C. Governmental Bureaucracy
   1. About 75 percent of the top-echelon positions in the federal government are held by white men.
   2. The governmental bureaucracy has been able to perpetuate itself and expand because it has many employees with highly specialized knowledge and skills who cannot easily be replaced by those from the "outside."
   3. The Iron Triangle and the Military Industrial Complex
      a. The *iron triangle* is a three-way arrangement in which a private interest group (usually a business corporation), a congressional committee or subcommittee, and a bureaucratic agency make the final decision on a political issue that is to be decided by that agency.
      b. A classic example of the iron triangle is an arrangement referred to as the **military-industrial complex** the mutual interdependence of the military establishment and private military contractors which started during World War II and continues to the present.

V. ECONOMIC SYSTEMS IN GLOBAL PERSPECTIVE
   A. The **economy** is the social institution that ensures the maintenance of society through the production, distribution, and consumption of goods and services.
   B. Pre-industrial, industrial, and postindustrial economies
      1. In pre-industrial economies, most workers engage in **primary sector production** -- the extraction of raw materials and natural resources from the environment.
      2. Industrialization brings sweeping changes to the economy as new forms of energy and machine technology proliferate as the primary means of producing goods and most workers engage in **secondary sector production** -- processing raw materials into finished goods.
      3. A postindustrial economy is based on **tertiary sector production** -- the provision of services (such as food service, transportation, communication, education, and entertainment) rather than goods.
   C. **Capitalism** is an economic system characterized by private ownership of the means of production, from which personal profits can be derived through market competition and without government intervention. "Ideal" capitalism has four distinctive features:
      1. Private ownership of the means of production;
      2. Pursuit of personal profit;
      3. Competition; and
      4. Lack of government intervention
   D. However, ideal capitalism does not exist in the U.S. for a number of reasons, including the presence of:
      1. **Oligopolies** -- where several companies control an entire industry.
      2. **Shared monopolies** -- where four or fewer companies supply 50 percent or more of a particular market.
      3. Mergers and acquisitions across industries create **conglomerates** -- combinations of businesses in different commercial areas, all owned by one holding company.

4. Competition also is reduced by **interlocking corporate directorates** -- members of the board of directors of one corporation who also sit on the board(s) of other corporations.
5. Government intervention often occurs in the form of regulations after some individuals and companies in pursuit of profits have run roughshod over weaker competitors; however, much "government intervention" has been in the form of aid to business.

E. **Socialism** is an economic system characterized by public ownership of the means of production, the pursuit of collective goals, and centralized decision making. "Ideal" socialism has three distinctive features:
1. Public ownership of the means of production;
2. Pursuit of collective goals; and
3. Centralized decision making.

F. A **mixed economy** combines elements of a market economy (capitalism) with elements of a command economy (socialism).
1. The term **democratic socialism** -- an economic and political system that combines private ownership of some of the means of production, governmental distribution of some essential goods and services, and free elections -- is sometimes used to refer to mixed economies.
2. The government in a mixed economy plays a larger role in setting rules, policies, and objectives.

G. A **Welfare state** exists when there is extensive government action to provide support and services to meet basic needs of its citizens.

VI. WORK IN THE CONTEMPORARY UNITED STATES

A. The economy in the United States and other contemporary societies is partially based on the work (purposeful activity, labor, or toil) that people perform. Professions -- high-status, knowledge-based occupations -- have five major characteristics.
1. Abstract, specialized knowledge;
2. Autonomy;
3. Self-regulation;
4. Authority; and
5. Altruism.

B. **Occupations** are categories of jobs that involve similar activities at different work sites.
1. The **primary labor market** is comprised of high paying jobs with good benefits that have some degree of security and the possibility for future advancement.
2. The **secondary labor market** is comprised of low-paying jobs with few benefits and very little job security or possibility for future advancement.

C. Upper Tier Jobs: Managers and Supervisors
1. Managers are essential in contemporary bureaucracies, where work is highly specialized and authority structures are hierarchical.
2. Workers at each level of the hierarchy take orders from their immediate supervisors and may give orders to a few subordinates.

D. Lower Tier or the Service Sector and Marginal Jobs

1. Positions in the lower tier of the service sector are part of the **secondary labor market**, characterized by low wages, little job security, few chances for advancement, higher unemployment, and very limited (if any) unemployment benefits.
2. Examples include janitors, waitresses, messengers, lower-level sales clerks, typists, file clerks, migrant laborers, and textile workers.
3. Many jobs in this sector are **marginal jobs**, which differ from the employment norms of the society in which they are located. These jobs lack regularity, stability, and adequate pay. They are often excluded from most labor legislation and typically provide no insurance or retirement benefits.
4. More than 11 million workers are employed in personal service industries, such as eating and drinking places, hotels, laundries, beauty shops, and household service, primarily maid service.

E. **Contingent work** is part-time work, temporary work, and subcontracted work that offers advantages to employers but often is detrimental to the welfare of workers.

1. Employers benefit by hiring workers on a part-time or temporary basis; they are able to cut costs, maximize profits, and have workers available only when they need them.
2. **Subcontracting** -- a form of economic organization in which a larger corporation contracts with other (usually smaller) firms to provide specialized components, product, or services -- is another form of contingent work that cuts employers' costs, but often at the expense of workers.

F. Unemployment

1. There are three major types of unemployment:
   a. *Cyclical unemployment* occurs as a result of lower rates of production during recessions in the business cycle.
   b. *Seasonal unemployment* results from shifts in the demand for workers based on conditions such as weather or seasonal demands such as holidays and summer vacations.
   c. *Structural unemployment* arises because the skills demanded by employers do not match the skills of the unemployed or because the unemployed do not live where the jobs are located.
2. The **unemployment rate** is the percentage of unemployed persons in the labor force actively seeking jobs. In 2004, the U.S. unemployment rate for white Americans was 4.9 percent; while for African Americans the rate was 10.5 percent.

G. Labor Unions and Worker Activism

1. U.S. labor unions have been credited with gaining an eight-hour work day, a five-day work week, health and retirement benefits, sick leave and unemployment insurance, and workplace health and safety standards for many employees through *collective bargaining* -- negotiations between labor union leaders and employers on behalf of workers.
2. Although the overall number of union members in the United States has increased since the 1960s, the proportion of all employees who are union

members has declined. Today about 17 percent of all U.S. employees belong to unions or employee associations.

3. Absenteeism and Sabotage
   a. Absenteeism is one means by which workers resist working conditions they consider to be oppressive.
   b. Other workers use sabotage to bring about informal work stoppages, such as "throwing a monkey wrench in the gears" to halt the movement of the assembly line.
   c. While most workers do not sabotage machinery, a significant number do resist what they perceive to be oppression from supervisors and employers.

H. Employment Opportunities for Persons with a Disability
   1. In 1990, the United States became the first nation to formally address the issue of equality for persons with a disability.
   2. Despite the passage of the Americans with Disability Act, about two-thirds of working-age persons with a disability in the U.S.A. are unemployed.

VII. POLITICS AND THE ECONOMY IN THE FUTURE
   A. Most futurists predict that transnational corporations will become even more significant in the global economy of tomorrow; while corporations will become even less aligned with the values of any one nation.
   B. Globally, millions of people remain committed to the ideals of democracy; in the United States, in spite of prejudice and discrimination, many ethnic and non-ethnic Americans continue to believe in a re-envisioning of democratic ideals.
   C. Politics and the economy are so intertwined in the United States and on a global basis that many social scientists speak of the two as a single entity: the *political economy*.
   D. The chasm between rich and poor nations will probably widen in the future as industrialized nations purchase fewer raw materials from industrializing countries and more products and services from one another.
   E. Around the world acts of terrorism extract a massive toll by producing rampant fear, widespread loss of human life, and extensive destruction of property.

ANALYZING AND UNDERSTANDING THE BOXES

After reading the chapter and studying the outline, re-read the boxes and write down key points and possible questions for class discussion.

Sociology and Everyday Life: How Much Do You Know About the Media?

Key Points:

Discussion Questions:

1.

2.

3.

You Can Make a Difference: Keeping an Eye on the Media

Key Points:

Discussion Questions:

1.

2.

3.

Sociology in a Global Perspective: Does Globalization Change the Nature of Social Policy?

Key Points:

Discussion Questions:

1.

2.

3.

PRACTICE TESTS

MULTIPLE CHOICE QUESTIONS

Select the response that best answers the question or completes the statement:

1. Free lance journalist Jill Carroll was taken hostage in: (p. 378)
   a. Iraq
   b. Iran
   c. Afghanistan
   d. North Korea

2. The social institution through which power is acquired and exercised by some people and groups is known as: (p. 380)
   a. government.
   b. politics.
   c. the state.
   d. the military.

3. __________ is the formal organization that has the legal and political authority to regulate the relationships among members of a society and between the society and those outside its borders. (p. 380)
   a. Economics
   b. Politics
   c. The military
   d. Government

4. According to the text, the state is: (p. 380)
   a. the social institution through which power is acquired and exercised by some people and groups.
   b. the formal organization that has the legal and political authority to regulate the relationships among members of a society and between the society and those outside its borders.
   c. the political entity that possesses a legitimate monopoly over the use of force within its territory to achieve its goals.
   d. the political entity that seeks to regulate all aspects of people's public and private lives.

5. ________ is the power that people accept as legitimate rather than coercive. (p. 382)
   a. Influence
   b. Clout
   c. Authority
   d. Legitimation

6. __________ is the ability of persons or groups to achieve their goals despite opposition from others. (p. 381)
   a. Influence
   b. Persuasiveness
   c. Power
   d. Charisma

7. According to Max Weber's classification of ideal types of authority, __________ is power that is legitimized on the basis of long-standing custom. (p. 383)
   a. traditional authority
   b. charismatic authority
   c. rational-legal authority
   d. (none of the above)

8. All of the following statements regarding racialized patriarchy are correct, *except*: (p. 383)
   a. Zillah R. Eisenstein coined the term "racialized patriarchy."
   b. Racialized patriarchy is closely intertwined with rational-legal authority.
   c. Racialized patriarchy is the continual interplay of race and gender.
   d. Racialized patriarchy reinforces traditional structures of power in contemporary societies.

9. According to Max Weber's classification of ideal types of authority, __________ is power that is legitimized on the basis of a leader's exceptional personal qualities or the demonstration of extraordinary insight and accomplishment that inspires loyalty and obedience from followers. (p. 383)
   a. traditional authority
   b. charismatic authority
   c. rational-legal authority
   d. (none of the above)

10. __________ tends to be temporary and relatively unstable because it derives primarily from individual leaders and a group of faithful followers. (p. 383)
    a. Traditional authority
    b. Charismatic authority
    c. Rational-legal authority
    d. (none of the above)

11. __________ occur(s) when charismatic authority is succeeded by a bureaucracy controlled by a rationally established authority or by a combination of traditional and bureaucratic authority. (p. 383)
    a. Traditional authority
    b. Routinization of charisma
    c. Rational-legal authority
    d. Primary groups

12. Rational-legal authority, also known as __________, is based on an organizational structure that includes a clearly defined division of labor, hierarchy of authority, formal rules, and impersonality. (p. 383)
    a. primary authority
    b. secondary authority
    c. monarchy
    d. bureaucratic authority

13. The National Socialist (Nazi) party in Germany during World War II is an example of a(n): (p. 386)
    a. monarchy.
    b. authoritarian regime.
    c. totalitarian regime.
    d. democracy.

14. The Taliban regime in Afghanistan is an example of a/an __________ political system. (p. 386)
    a. authoritarianism
    b. totalitarianism
    c. dictatorship
    d. military junta

15. All of the following statements regarding democracy are true, except: (pp. 386-387)
    a. Democracy is a political system in which the people hold the ruling power either directly or through elected representatives.
    b. Several nations have attempted direct democracy at the national level.
    c. Representative democracy is not always equally accessible to all people in a nation.
    d. The framers of the Constitution established a system of representative democracy in the United States.

16. Political action committees: (p. 389)
    a. generally have been abolished by reforms in campaign finance laws.
    b. are comprised of people who volunteer their time but not money to political candidates and parties.
    c. are organizations of special interest groups that fund campaigns to help elect candidates based on their stances on specific issues.
    d. encourage widespread political participation by citizens at the grassroots level.

17. The power elite model was developed by: (pp.389-390)
    a. Emile Durkheim.
    b. Thomas R. Dye and Harmon Zeigler.
    c. Joseph Steffan.
    d. C. Wright Mills.

18. The organizations responsible for developing and articulating policy positions, educating voters about issues, and recruiting candidates to run for office are known as: (p. 391)
    a. political parties.
    b. political action committees.
    c. interest groups.
    d. federal election committees.

19. The three-way arrangement in which a private interest group, a congressional committee, and a bureaucratic agency make the final decision on a political issue that is to be decided by that agency is known as: (pp. 393-394)
    a. political subversion.
    b. the iron law of oligarchy.
    c. the iron triangle of power.
    d. the power elite.

20. All of the following are examples of primary sector production, *except*: (pp. 394-395)
    a. oil wells
    b. coal mining
    c. metal ore production
    d. logging

21. Industrial economies are characterized by _______ sector production. (p. 395)
    a. primary
    b. secondary
    c. tertiary
    d. quartiary

22. According to the text, all of the following are features of capitalism, *except*: (pp. 396-400)
    a. private ownership of the means of production.
    b. pursuit of personal profit.
    c. competition.
    d. governmental intervention in the marketplace.

23. If Sony, EMI, Time Warner, and only a few other companies control the entire music industry, this arrangement would be a(n): (p. 397)
    a. oligopoly.
    b. monopoly.
    c. conglomerate.
    d. amalgamation.

24. ______ is an economic system characterized by public ownership of the means of production, the pursuit of collective goals, and centralized decision making. (p. 396)
    a. Capitalism
    b. Communism
    c. Socialism
    d. Communitarianism

25. Democratic socialism is characterized by: (p. 402)
    a. extensive government action to provide support and services to its citizens.
    b. private ownership of the means of production, from which personal profits can be derived through market competition and without government intervention.
    c. public ownership of the means of production, the pursuit of collective goals, and centralized decision making
    d. private ownership of some of the means of production, governmental distribution of some essential goods and services, and free elections.

26. _____ unemployment results from shifts in the demand for workers based on conditions such as the weather or the season. (p. 406)
    a. Inclement
    b. Occasional
    c. Seasonal
    d. Cyclical

27. _____ refers to negotiations between employers and labor union leaders on behalf of workers. (p. 407)
    a. Collective bargaining
    b. Mediation
    c. Binding arbitration
    d. Agency shop tactics

28. The phrase "throwing a monkey wrench in the gears" originated with the practice of: (p. 408)
    a. managers deliberately slowing down the assembly line.
    b. workers who poke fun at their job by dressing up like monkeys.
    c. lower tier employees trying to impress their managers.
    d. "losing" a tool in assembly line machinery.

29. In 1990, __________ became the first nation to formally address the issue of equality for persons with a disability (p. 378)
    a. the United States
    b. Sweden
    c. Norway
    d. Japan

30. What do NAFTA, EU, and GATT have in common? (p. 378)
    a. They are the most powerful labor unions in the U.S.A.
    b. They are the most powerful labor unions in Europe
    c. They are all examples of transnational trade agreements
    d. They are transnational trade agreements founded by Asian nations.

TRUE-FALSE QUESTIONS

T F 1. Political sociology is the same as political science. (p. 380)

T F 2. Power is a social relationship that involves both leaders and followers. (p. 381)

T F 3. Charismatic authority tends to be temporary and relatively unstable. (p. 383)

T F 4. A nation-state is a unit of political organization that has recognizable national boundaries and whose citizens possess specific legal rights and obligations. (p. 385)

T F 5. In a representative democracy, elected representatives are expected to keep the "big picture" in mind, and they do not necessarily need to convey the concerns of those they represent in all situations. (p. 387)

T F 6. Functionalists suggest that divergent viewpoints lead to a system of political pluralism in which the government functions as an arbiter between competing interests and viewpoints. (p. 387)

T F 7. Over the past two decades, special interest groups have become more involved in multiple-issue politics because so many political issues are intertwined with other concerns. (p. 389)

T F 8. Sociologist G. William Domhoff believes that the ruling class wields sufficient power to constrain political processes in the United States. (p. 390)

T F 9. The United States has one of the highest percentages of voter turnout of all Western nations. (p. 392)

T F 10. The economy is the social institution responsible for the production, distribution, and consumption of goods and services. (p. 394)

T F 11. Preindustrial economies typically are based on the extraction of raw materials and natural resources from the environment. (pp. 394-395)

T F 12. A conglomerate exists when four or fewer companies supply 50 percent or more of a particular market. (p. 398)

T F 13. According to Karl Marx, socialism and communism are virtually identical. (p. 401)

T F 14. Children whose parents are professionals are more likely to become professionals themselves. (p. 403)

T F 15. Temporary workers are the fastest growing segment of the contingent work force. (p. 406)

SOCIOLOGY IN OUR TIMES: DIVERSITY ISSUES

1. Do you think that you underrate the significance of politics in your daily life? Why? If so, what are you doing to ensure your own well-being in our society?

2. Have labor unions historically contributed to the growth of a middle class in the United States? If labor unions decline in the future, will this have any impact on the size of the middle class in the future?

3. According to the text, "race and gender are factors in access to the professions." (p. 434) Do you agree or disagree with this statement? Can you give examples from your personal experience that tend to confirm or deny that statement?

4. Why are large numbers of young people, people of color, recent immigrants, and white women over represented in marginal jobs and other positions in the lower tier of the service sector? Do employed college students sometimes see themselves as temporarily "stuck" in jobs in their tier while many other workers see themselves as permanently "stuck" in such positions?

5. How does resistance help people in lower-tier service jobs survive at work? (p. 439) Have you ever engaged in your own form of absenteeism, sabotage, or resistance at school or work?

6. What is the purpose of the Americans with Disabilities Act? Do you think this law is effective? Why? Why not?

7. Do you think all workplaces of the future will reflect the increasing diversity currently occurring in the U.S. population? In your future employment, will you need to communicate with individuals whose race/ethnicity, gender, and class background may be different from your own? Are you taking steps to prepare for employment in a multicultural or global workplace? If so, how?

INTERNET EXERCISES

Use a search engine to assist you completing the following:

1. Choose a company online and read their disability benefits. Then, explain the qualifications for receiving those benefits.

2. Find five labor unions and explain how these unions protect employees

INFOTRAC COLLEGE EDITION ONLINE READINGS AND EXERCISES

Access the Infotrac Online Library to answer the following. Log on and enter "It's a Different World for Lawmakers of 2025" into the **keyword** search field and hit return. Click to read the article and then answer the following questions:.

a. What four different U.S.A. generations comprise the political agenda today?

b. What are major political and technological changes?

Return to the search field. Next type "Anti-Capitalism and the terrain of social justice " into the field and hit return. Read the article and answer the following questions:

a. Explain anti-capitalism, according to the article.

b. What were the people in Seattle protesting against?

STUDENT CLASS PROJECTS AND ACTIVITIES

1. The text notes the low rate of voter participation in our country. You are to attempt to discover some explanations for this condition. You are to conduct an informal survey of a cross section of 20 people of different ages, ethnicity, social class, race, and sex. On your survey forms, you must include the basic background information of your respondents (name, race, sex, address, age, ethnicity, occupation, or student status). Obtain responses to the following questions: (1) When was the last time you voted? Why? (2) Did you vote for president in the last election? (3) Do you usually vote in all elections? Why? Why not? (4) Do you actually vote along political party lines, such as for the Democrats or Republicans? If so, which party? Why? (5) Do other members of your family vote? (6) Do you know at least one elected official? Who? (7) Do you support the present position of the President of the United States on most issues? Explain. Formulate any other question you may want to ask. Provide a conclusion, an analysis, and an evaluation of the data in writing up this project. Submit your paper at the appropriate time.

2. Trace the history (including the origin and the growth) of labor unions in the United States, for both male and female workers. Include in your research a statement of the present status of labor unions in the present American marketplace.

3. Trace the growth, development, and change of blue-collar and white-collar jobs in the American work force. Include in your research the present status of blue-collar versus white-collar jobs. Some possible sources are: the Economic Policy Institute (a think-tank in Washington D.C.) and government documents such as the Historical Statistics of the United States. Include, when feasible, any charts and graphs, and provide a summary and conclusion of your research.

ANSWERS TO PRACTICE TESTS FOR CHAPTER 13

Answers to Multiple Choice Questions

1. a — Jill Carroll, a freelance journalist on assignment for the Christian Science Monitor was taken hostage in Iraq in 2005.. (p. 378)
2. b — The social institution through which power is acquired and exercised by some people and groups is known as politics. (p.380)
3. d — The government has the legal authority and in most cases the popular authority. (p. 380)
4. c — According to the text, the state is the political entity that possesses a legitimate monopoly over the use of force within its territory to achieve its goals. (p. 380)
5. c — Authority is the power that people accept as legitimate rather than coercive. (p. 382)
6. c — The important point here is despite opposition. (p. 381)
7. a — We find this in older countries that have long standing traditions and authority associated with them. (p. 383)
8. b — All of the following statements regarding racialized patriarchy are correct, *except*: racialized patriarchy is closely intertwined with rational legal authority. Instead, racialized patriarchy is closely intertwined with traditional authority. (p. 383)
9. b — Often times associated with revolutions or religious leaders. (p. 383)
10. b — And the leaders only last for a short time, and their replacements typically do not have the charisma to sustain the authority. (p. 383)
11. b — If the leaders are able to sustain the authority after the charismatic leader you have this. (p. 383)
12 d — The cornerstone of modern day corporations. (p. 383)
13. c — The National Socialist (Nazi) party in Germany during World War II is an example of a totalitarian regime. (p. 386)
14. b — The Taliban regime maintained absolute control over the Afghan people in most of that country. (p. 386)
15. b — All of the following statements regarding democracy are true, *except*: several nations have attempted direct democracy at the national level. (pp. 386-387)
16. c — Political action committees are organizations of special interest groups that fund campaigns to help elect candidates based on their stances on specific issues. (p. 389)
17. d — The power elite model was developed by C. Wright Mills. (pp. 389-390)
18. a — The organizations responsible for developing and articulating policy positions, educating voters about issues, and recruiting candidates to run for office are known as political parties. (p. 391)
19. c — The three-way arrangement in which a private interest group, a congressional committee, and a bureaucratic agency make the final decision on a political issue that is to be decided by that agency is known as the iron triangle of power. (pp. 393-394)
20. c — Metal ore production is an example of *secondary*, not *primary* sector production. (pp. 394-395)

21. b Industrial economies are characterized by secondary sector production. (p. 395)
22. d According to the text, all of the following are features of capitalism except: governmental intervention in the marketplace. (pp. 396-400)
23. a If Sony, EMI, Time Warner and only a few other companies control the entire music industry, this arrangement would be an oligopoly. (p. 397)
24. c Socialism is an economic system characterized by public ownership of the means of production, the pursuit of collective goals, and centralized decision making. (p. 396)
25. d Democratic socialism is characterized by private ownership of some of the means of production, governmental distribution of some essential goods and services, and free elections. (p. 402)
26. c Seasonal unemployment results from shifts in the demand for workers based on conditions such as the weather or the season. (p. 406)
27. a Collective bargaining refers to negotiations between employers and labor unions on behalf of workers. (p. 407)
28 d the phrase "throwing a monkey wrench in the gears" originated with the practice intentionally "losing" a tool in assembly line machinery. (p. 408)
29. a In 1990, the United States passed the Americans with Disabilities Act (p. 408)
30. c NAFTA, the EU, and GATT are examples of transnational trade agreements. (p. 408)

Answers to True-False Questions

1. False Political sociology focuses on the social circumstances of politics and explores the interrelationships between politics and social structures. Political science primarily focuses on power and its distribution in different types of political systems. (p. 380)

2. True (p. 381)

3. True (p.383)

4. True (p.385)

5. False In a representative democracy, elected representatives are expected convey the concerns and interests of those they represent. (p. 387)

6. True (p. 387)

7. False Over the past two decades, special interest groups have become more involved in single issue politics. (p.389)

8. True (p. 390)

9. False The United States has one of the lowest percentages of voter turnout of all Western nations. (p. 392)

10. True (p. 394)

11. True (pp. 394-395)

12. False A shared monopoly exists when four or fewer companies supply 50 percent or more of a particular market. (p. 398)

13. False According to Karl Marx, communism is an economic system characterized by common ownership of all economic resources. He believed that socialism was merely a temporary stage en route to an ideal communist society. (p. 401)

14. True (p. 403)

15. True (p. 406)

CHAPTER 14
HEALTH, HEALTH CARE, AND DISABILITY

CHAPTER SUMMARY

Although many people may think of health as simply being the absence of disease, the World Health Organization defines **health** as a state of complete physical, mental and social well-being. **Health care** is any activity intended to improve health. The **infant mortality rate,** the number of deaths of infants under one year of age per 1,000 live births in a given year, and **life expectancy,** the estimate of the average lifetime of people born in a specific year, are two indicators of the well-being of people in a society. A vital part of health care is **medicine,** an institutionalized system for the scientific diagnosis, treatment, and prevention of illness. In analyzing health from a global perspective, we find that health and health care expenditures vary widely among nations. **Social epidemiology** is the study of the causes and distribution of health, disease, and impairment throughout a population. Factors such as age, sex, race/ethnicity, social class, lifestyle choices, and drug use and abuse affect the health and longevity of people in a given area. Sexual activity, usually associated with good physical and mental health, may involve health-related hazards, such as transmission of sexually transmitted diseases. The fee-for-service method continues in the United States, which is expensive; few restrictions are placed on medical fees. The United States and the Union of South Africa are the only developed nations without some form of universal health coverage for all citizens. Private health insurance, public health insurance—consisting of Medicare and Medicaid—health maintenance organizations, and managed care compromise the major approaches for paying for U.S. health care. Advances in high-tech medicine is becoming a major part of overall health care; however, many people are turning to holistic medicine -- is an approach to health care that focuses on prevention of illness and disease and is aimed at treating the whole person-body and mind-rather than just the part or parts in which symptoms occur. Many practitioners of **alternative medicine**-healing practices inconsistent with dominant medical practice- take a holistic approach. According to a **functionalist** approach, if a society is to function as a stable system, people must be healthy in order to contribute to their society. People who assume the **sick role** are unable to fulfill their necessary social roles, contributing to an inability to fulfill their functions. Unequal access to good health and health care is a major issue of concern to **conflict** theorists. The **medical-industrial complex**-that which encompasses both local physicians and hospitals as well as global health-related industries- produces and sells medicine as a commodity; people below the poverty level and those just above it have great difficulty gaining access to medical care. **Symbolic Interactionists**, in explaining the social construction of illness, focus on the meaning that social actors give their illness or disease and how that meaning affects their self-concept and their relationships with others. According to a postmodernist approach, doctors have gained power through the clinical gaze, and gained prestige through the classification of disease. **Disability**, a physical or health condition that stigmatizes or causes discrimination is often defined in terms of work from a business perspective, or in terms of organically based impairments from a medical professional perspective. Health care in the future is critically affected by health care technologies; however, health in the future will somewhat be up to each of us- how we safeguard ourselves against illness, and how we help others who are victims of diseases and disabilities.

LEARNING OBJECTIVES

After reading chapter 14, you should be able to:

1. Define health, health care and disability and explain the importance of these issues for individuals the whole of society.

2. Compare health in a global perspective to health in the United States.

3. Define social epidemiology and its role in society.

4. Compare and contrast how age, sex, race/ethnicity and social class affect health and mortality.

5. Explain how lifestyle choices affect health, disease, and impairment.

6. Describe the rise of scientific medicine in the United States.

7. Explain four methods of paying for health care and controlling health care costs in the United States.

8. List and discuss three major social implications of advanced medical technology.

9. Compare and contrast holistic medicine and alternative medicine with traditional, or orthodox medical treatment.

10. Distinguish between health care in the United States to that of other countries.

11. Describe functionalist, conflict, symbolic interactionist, and postmodernist perspectives on health and medicine.

12. Explain the sociological perspectives on disability.

13. Discuss inequalities related to disability and describe the impact of stereotypes, prejudice, and discrimination on persons with disabilities.

14. Describe and discuss some options for health care in the future.

KEY TERMS (defined at page number shown and in glossary)

acute diseases 420
chronic diseases 420
demedicalization 435
disability 436
drug 421
health 415
health care 416
health maintenance organization 427
holistic medicine 430
infant mortality rate 417
life expectancy 416
managed care 427
medical-industrial complex 434
medicalization 434
medicine 416
sick role 431
social epidemiology 420
socialized medicine 428
universal health care 428

CHAPTER OUTLINE

I. HEALTH IN GLOBAL PERSPECTIVE
   A. The World Health Organization defines **health** as a state of complete physical, mental and social well-being.
   B. Illness refers to an interference with health.
   C. **Health care** is any activity intended to improve health.
   D. **Medicine**: an institutionalized system for the scientific diagnosis, treatment, and prevention of illness; it is a vital part of health care.
   E. Some measures of health and well-being are:

1. **Life expectancy**: the estimate of the average lifetime of a people born in a specific year.
2. The **infant mortality rate**: the number of deaths of infants under one year of age per 1,000 live births in a given year.

F. Health care and health care expenditures vary widely among nations; the WHO reports that babies born in low-income countries will not live anywhere near as long as children in high-income countries.

II. HEALTH IN THE UNITED STATES

A. **Social epidemiology**, the study of the causes and distribution of health, disease, and impairment throughout a population, provides some explanations of health and disease.

B. Social epidemiologists investigate disease agents, the environment, and the human host as sources of illness.

C. Younger people are more likely to have **acute diseases**, illnesses that strike suddenly and cause dramatic incapacitation and sometimes death, while older people are more likely to have **chronic diseases**, illnesses that are long-term and develop gradually or are present from birth.

D. Today, women on average live longer than men; at birth, females have lower mortality rates both in the prenatal stage and during the first month of life.

E. Some research points out that social class is a greater determinant of health than race/ethnicity; however, illness is also related to a person's occupation. An exception to this claim appears to be Latinas/os, who are more likely to be poorer than non-Latinas/os whites, but yet have an overall lower death rate from heart disease, cancer, accidents, and suicide.

F. Three (of many) *lifestyle factors* relate to health:

1. **Drugs**, any substances -- other than food and water -- that, when taken into the body, alters its functioning in some way, affect health in many ways.
   a. Long-term heavy use of *alcohol* can damage the brain and other parts of the body, cause nutritional deficiencies, cardiovascular problems, and alcoholic cirrhosis.
   b. *Nicotine* (tobacco) is a toxic, dependency-producing psychoactive drug that is more addictive than heroin; smoking tobacco is linked to cancer and other serious diseases. *Environmental tobacco smoke-* -- the smoke in the air that nonsmokers inhale from other people's tobacco smoking -- is hazardous for nonsmokers.
   c. The use of *illegal drugs*, like cocaine and marijuana, affect the lifestyle and the health of individuals.
2. Sexual activity may involve health-related hazards such as transmission of certain *sexually transmitted diseases.*

a. The most prevalent STDs are HIV/AIDS, gonorrhea, syphilis, and genital herpes. All of these affect the health of individuals; however, the most serious of these is AIDS.

b. On a global basis, the number people infected with HIV or AIDS is increasing; fourteen percent of all new cases worldwide are children; more than two-thirds of all people with HIV/AIDS live in sub-Saharan Africa; fourteen percent live in south and southeast Asia.

c. HIV is transmitted through unprotected sexual intercourse with an infected partner, by sharing a hypodermic needle with someone who is infected, by exposure to contaminated blood or blood products, and by an infected woman passing the virus on to a child during pregnancy, childbirth, or breast feeding.

3. Staying healthy: diet and exercise

    a. Lifestyle choices can include positive actions such as maintaining a healthy diet and good exercise program.

III. HEALTH CARE IN THE UNITED STATES

A. The rise of scientific medicine resulted from several significant discoveries during the nineteenth century in areas such as bacteriology and anesthesiology; at the same time, advocates of reform in medical education were promoting scientific medicine.

1. The Flexner Report, complied by Abraham Flexner, included a model of medical education; those medical schools that did not fit the "model" were closed. Only two of the African American medical schools survived; and only one of the medical schools for women survived.

2. With professionalization resulting from the Flexner Report, licensed medical doctors gained control over the entire medical establishment; this continues today -- and may continue into the future.

B. Medicine Today

1. Paying for health care in the United States has traditionally been on a fee-for-service basis.

2. Patients are billed individually for each service they receive.

    a. This includes treatment by doctors, laboratory work, hospital visits, other health-related expenses, and prescriptions.

    b. This payment is expensive; few restrictions are placed on the fees that doctors, hospitals, and medical providers can charge.

    c. However, this method has led to advances in medicine because of monies provided for innovations and techniques.

C. Paying for medical care in the United States

1. Private health insurance companies escalated rapidly, beginning in the 1960s.
   a. Third party providers began to pay large portions of doctor and hospital bills for insured patients, which led to medical inflation.
   b. Patients have no incentives to limit their visits to doctors or to hospitals because they have already paid their premiums and feel entitled to medical care, regardless of the cost.
2. Public health insurance is subsidized by federal and state taxes; the U.S. has two nationwide public health insurance programs: Medicare, for those 65 or older who are eligible; and Medicaid, a jointly funded federal-state-local program to make health care more available to the poor. Both programs are in financial difficulty today.
3. **Health Maintenance Organizations (HMOs)** were created to provide workers with health coverage by keeping costs down; HMOs provide, for a set monthly fee, total care with an emphasis on prevention to avoid costly treatment later.
   a. Research shows that preventative care is good for the individual's health but does not necessarily lower total costs.
   b. Critics charge that those HMOs whose primary-care physicians are paid on a captation basis – meaning that they receive only a fixed amount per patient that they see, regardless of how long they spend with that patient – in effect encourage doctors to under-treat patients.
4. **Managed care** is any system of cost containment that closely monitors and controls health care providers' decisions about medical procedures, diagnostic tests, and other services that should be provided to patients.
   a. Patients choose a primary-care physician from a list of participating doctors.
   b. Doctors must get approval before they perform any procedures or admit a patient to a hospital; if they fail to obtain advance approval, the insurance company has the right to refuse to pay for the treatment or the hospital stay.
5. The *uninsured and the underinsured* comprise one-third of all U.S. citizens. The working poor constitutes a large portion; they make too little to afford health insurance, too much to qualify for Medicaid, and their employers do not provide health insurance.

D. Paying for medical care in other nations

1. Canada has had a **universal health care system** since 1972, a health care system in which all citizens receive medical services paid for by tax revenues; in Canada, these revenues are supplemented by insurance premiums paid by all taxpaying citizens. It does not

constitute what is referred to as **socialized medicine**, a health care system in which the government owns the medical care facilities and employs the physicians; the physicians are not government employees.

2. Great Britain, in 1946, passed the National Health Service Act, which provided for all health care services to be available at no charge to the entire population. The government sets health care policies, raises funds and controls the medical care budget, owns health care facilities, and directly employs physicians and other health personnel.
   a. This health care system does constitute socialized medicine.
   b. Physicians may accept private patients who pay for their own care; these patients are just a small fraction of a physician's practice.
3. China adopted innovative strategies after a lengthy civil war in order to improve the health of its populace; physician extenders, known as *street doctors* in urban areas and *barefoot doctors* in rural areas, had little formal training and worked under the supervision of trained physicians.
   a. Today, medical training is more rigorous; all doctors receive training in both Western and traditional Chinese medicine.
   b. Doctors who work in hospitals receive a salary; all other doctors are on a fee-for-service basis. Chinese who can afford it purchase health care insurance to cover the rising costs of hospitalization. The health of China's citizens is only slightly below that of most industrialized nations.

E. Social implications of advanced medical technology include:
   1. The new technologies create options for people and society, but options that alter human relationships.
   2. The new technologies increase the cost of medical care (such as the computerized axial tomography, CT or CAT scanner).
   3. The new technologies raise provocative questions about the very nature of life, such as duplicating mammals from adult DNA.

F. Holistic medicine and alternative medicine provide variations from traditional (or orthodox) medical treatment
   1. **Holistic medicine** focuses on prevention of illness and disease and aims at treating the whole person -- body and mind -- rather than just the part or parts in which symptoms occur.
   2. *Alternative medicine* involves healing practices inconsistent with dominant medical practice, which takes a holistic approach; many people today are utilizing this in addition to or in lieu of traditional medicine.

IV. SOCIOLOGICAL PERSPECTIVES ON HEALTH AND MEDICINE

A. The functionalist perspective on health: the sick role

1. According to Talcott Parsons, the **sick role** is the set of patterned expectations that define the norms and values appropriate for individuals who are sick and for those who interact with them.
2. The sick role has four characteristics:
   a. A person's sickness is not deliberate;
   b. A sick person is exempted from responsibilities;
   c. A sick person must want to get well; and
   d. A sick person must seek competent help from a medical professional.
3. Illness is dysfunctional for individuals and the larger society; sick people are unable to fulfill their necessary social roles; it is important for the society to maintain social control over people who enter the sick role.

B. The conflict perspective: inequalities in health and health care

1. Conflict theorists emphasize the political, economic, and social forces that affect health and the health care delivery system.
2. Medicine is a commodity that is produced and sold by the **medical-industrial complex**, that which encompasses both local physicians and hospitals as well as global health-related industries such as insurance companies and pharmaceutical and medical supply companies that deliver health care today.
3. Medical care is linked to people's ability to pay and their position within the class structure; the affluent or those who have good medical insurance most likely receive care; the *medically indigent*, those who cannot qualify for Medicaid, but do not earn enough cannot afford private medical care.
4. Physicians hold a legal monopoly over medicine; they can charge inflated fees; clinics, pharmacies, laboratories, hospitals, insurance companies, and many other corporations derive excessive profits from the existing system of payment in medicine.

C. Symbolic interactionist perspective: the social construction of illness

1. Symbolic interactionists focus on the meaning that actors give their illness or disease and how this will affect their self-concept and their relationship with others.
2. In addition to the objective criteria for determining medical conditions, the subjective component is very important; the term **medicalization** refers to the process whereby nonmedical problems become defined and treated as illness or disorders.
3. Medicalization may occur on three levels: (a) the conceptual level; (b) the institutional level; and (c) the interactional level. Medicalization is typically the result of a lengthy promotional

campaign, often culminating in legislation or other social policy changes that institutionalize a medical treatment of a new "disease."
4. **Demedicalization** refers to the process whereby a problem ceases to be defined as an illness or a disorder.
5. Symbolic interactionists provide new insights on how illness may be socially constructed, and not strictly determined by medical criteria.

D. A Postmodernist perspective: The Clinical Gaze
1. According to Michel Foucault, the formation of clinical medicine led to the power that doctors gained over other medical personnel and everyday people.
2. Doctors gain power through the *clinical gaze,* which they use to gather information.
    a. As doctors "diagnose", they become experts.
    b. As a definitive network of disease classification, and new rules and new tests developed, the dominance of the doctor's wisdom became enhanced.

V. DISABILITY

A. **Disability**, refers to a reduced ability to perform tasks one would normally do at a given stage of life and that may result in stigmatization or discrimination against a person with disabilities.

B. An estimated 49.7 million persons in the U.S. have one or more physical or mental disabilities, and the number is increasing as medical advances make it possible for those who would have died from an accident or illness to survive, but with an impairment, and as life expectancies increase.

C. Environment, lifestyle, and working conditions may contribute to disability.

D. Many disability right advocates argue that persons with a disability are kept out of the mainstream of society, being denied equal opportunities in education by being consigned to special education classes or schools.

E. When confronted with a disability, most people adopt one of two strategies- avoidance or vigilance.

F. Sociological Perspectives on Disability
1. Functionalist Talcott Parsons focused on how people who are disabled fill the sick role. This is a *medical model* of disability; people with disabilities become chronic patients.
2. Symbolic interactions examine how people are labeled as a result of disability; Eliot Freidson determined that the particular label results from: (1) the person's degree of responsibility for the impairment, (2) the apparent seriousness of the condition, and (3) the perceived legitimacy of the condition.

3. From a conflict perspective, persons with disabilities are a subordinate group in conflict with persons in positions of power in the government, the health care industry, and the rehabilitation business. When people with disabilities are defined as a social problem and public funds are spent to purchase goods and services for them, rehabilitation becomes big business.

G. Social inequalities based on disability include prejudice and discrimination.

1. Stereotypes of persons with a disability fall into two categories: (1) deformed individuals who also may be horrible deviants; or (2) persons who are to be pitied. Even positive stereotypes become harmful to people with a disability, those that excel despite the impairment. Disability rights advocates note that such stereotypes do not reflect the daily struggle of most people with disabilities.
2. Employment, poverty, and disability are related; people may become economically disadvantages as a result of chronic illness or disability; on the other hand, poor people are less likely to be educated and more likely to be malnourished and have inadequate access to health care.
3. Disability has a stronger negative effect on women's labor participation than it does on men's.

VI. HEALTH CARE IN THE FUTURE

A. In the future, advanced health care technologies for high income countries will provide more accurate and quicker diagnosis, effective treatment techniques, and increased life expectancy. However, there will also be new ethical concerns: the inability of the poor of the world, (including the U. S.) to have access to health care, the growth of managed-care companies in the U. S. making decisions formerly reserved for physicians and hospital personnel and ethical concerns relating to technological advances.

B. Ultimately, health in the future will somewhat be up to each of us -- what measures we take to safeguard ourselves against illness and disorders, and how we help others who are victims of acute and chronic diseases or disabilities.

ANALYZING AND UNDERSTANDING THE BOXES

After reading the chapter and studying the outline, re-read the boxes and write down key points and possible questions for class discussion.

Sociology and Everyday Life: How Much Do You Know About Health, Illness, and Health Care?

Key Points:

Discussion Questions:

1.

2.

3.

Framing Health Issues: Death and Dying as a Television Spectacle

Key Points:

Discussion Questions:

1.

2.

3.

You Can Make a Difference: Joining The Fight Against Illness and Disease!

Key Points:

Discussion Questions:

1.

2.

3.

PRACTICE TESTS

MULTIPLE CHOICE QUESTIONS

Select the response that best answers the question or completes the sentence:

1. The World Health Organization (WHO) defines health as: (pp. 415-416)
   a. the absence of disease.
   b. complete physical, mental, and social well-being.
   c. the body in a state of equilibrium, with all parts in balance.
   d. the absence of sickness, viruses, and pains

2. One aspect of institutional healing is health care and the _____ system in a society. (p. 416)
   a. health care delivery
   b. medical model
   c. hospital
   d. clinical

3. _____ refers to the positive sense of complete well being; while______ refers to an interference with health. (p. 416)
   a. healing; disease
   b. health; healing
   c. health; illness
   d. health care; disease

4. Health care: (p. 416)
   a. is positively linked to money spent on health care and people's physical, mental, and social well-being.
   b. equates positively with longevity.
   c. is more expensive, but less productive, for individuals in Sweden than those in the United States.
   d. is any activity intended to improve health.

5. _____ is an institutionalized system for the scientific diagnosis, treatment, and prevention of illness. (p. 416)
   a. The medical-industrial complex
   b. Universal health care
   c. Medicine
   d. Disease

6. The number of people infected with HIV/AIDS __________ between 1990 and 2000. (p. 416)
   a. dropped by half
   b. remained about the same
   c. more than doubled
   d. increased by a factor of 4

7. _________ refers to an estimate of the average lifetime of people born in a specific year. (pp. 416-417)
   a. The demographic transition
   b. Demography
   c. The infant mortality rate
   d. Life expectancy

8. In the low-income nation of Zambia, approximately _____ percent of the people are not expected to live to age 60. (p. 417)
   a. 20
   b. 40
   c. 60
   d. 80

9. The World Health Organization estimates that about _____ of the babies born in low-income countries during 2004 will die during the _____. (p. 418)
   a. one-third; first year of life.
   b. one-third; first month of life.
   c. two-thirds; first year of life.
   d. two-thirds; first month of life.

10. The United States spends the equivalent of _____ per person on health care each year. (p. 419)
    a. $385
    b. $3,925
    c. $38,000
    d. $380,000

11. Sweden spends less per person on health care each year than does the United States. Average life expectancy in Sweden is __________ the average life expectancy in the U.S. (p. 419)
    a. lower than
    b. about the same as
    c. higher than
    d. (no data are available to determine this)

12. Social epidemiologists are people who study the causes and distribution of: (p. 420)
    a. health, disease, and impairment in a population.
    b. epidemics that occur throughout a population.
    c. mental illness and behavior disorders in a population.
    d. the role of degenerative diseases in a population.

13. _________ is the study of the causes and distribution of health, disease, and impairment throughout a population. (p. 420)
    a. Medicine
    b. Health care
    c. Population demography
    d. Social Epidemiology

14. __________ are illnesses that are long term or lifelong and that develop gradually or are present from birth. (p. 420)
    a. Acute diseases
    b. Chronic diseases
    c. Ascribed conditions
    d. Achieved conditions

15. ______are illnesses that strike suddenly and cause dramatic incapacitation and sometimes death are called: (p. 420)
    a. Chronic diseases
    b. Disabilities
    c. Epidemics
    d. Acute diseases

16. _______are some serious consequences of alcoholism. (p. 422)
    a. Alcoholic cirrhosis
    b. Nutritional deficiencies
    c. Cardiovascular problems
    d. all of the above
    e. both “a” and “b”, but not "c"

17. ____ is a toxic, dependency producing drug that is more addictive than ______. (p. 422)
    a. Heroin; nicotine
    b. Nicotine; heroin
    c. Nicotine; marijuana
    d. Marijuana; alcohol

18. Until the 1960s, ____ and _____ were the principal STDs in this country. (p. 423)
    a. genital herpes; syphilis
    b. gonorrhea; genital herpes
    c. gonorrhea; syphilis
    d. genital herpes; HIV/AIDS

19. More than two-thirds percent of all people with HIV/AIDS live in: (p. 423)
    a. sub-Saharan Africa.
    b. South Asia.
    c. Southeast Asia.
    d. the United States.

20. The Flexner Report stated in 1910 that: (p. 424)
    a. new medical schools should be developed for women.
    b. more medical schools should be developed for African Americans.
    c. new medical schools should be developed in rural areas.
    d. most existing medical schools were inadequate.

21. Throughout most of the twentieth century, medical care in the United States was paid for: (p. 425)
    a. by HMOs.
    b. on a fee-for-service basis.
    c. by third party providers.
    d. by managed care.

22. ______is a jointly funded federal-state local health care program for the poor. (p. 427)
    a. Medicare
    b. Medical
    c. Medicaid
    d. Managed care

23. The only high-income nations without some form of universal health coverage for all its citizens are: (p. 426)
    a. the United States and the Union of South Africa.
    b. Canada and the United States.
    c. Japan and the United States.
    d. Great Britain and the United States.

24. An approach to health care that treats the whole person rather than just symptoms that occur is known as: (p. 430)
    a. mainstream medicine.
    b. managed care.
    c. holistic medicine.
    d. medicalization.

25. Individuals who had little medical formal training in China and who work in rural communities, emphasizing health and health care and treating illness and disease, are known as: (p. 429)
    a. barefoot doctors.
    b. Red Guard doctors.
    c. street doctors.
    d. communist herb doctors.

26. People in the sick role are expected to: (p. 431)
    a. be responsible for their condition.
    b. desire to get well.
    c. continue their normal roles and obligations.
    d. ignore competent medical assistance

27. __________ refers to the process whereby nonmedical problems become defined and treated as illnesses or disorders. (pp. 434-435)
    a. Treatment
    b. Therapy
    c. Medical definition
    d. Medicalization

28. Symbolic interactionist theorists note that defining practices such as gambling as a psychological illness ("compulsive gambling") allows physicians to determine what is "normal" and "acceptable" behavior. This process is known as __________. (p. 435)
    a. medical definition
    b. psychological definition
    c. medicalization of deviance
    d. psychological deviance

29. The American Psychiatric Association has removed homosexuality from its list of mental disorders. This process of removal is referred to by sociologists as __________. (p. 435)
    a. medicalization of deviance
    b. demedicalization
    c. psychiatric removal
    d. demystification

30. Women with disabilities are _______ to be covered by pension and health plans than are men. (p. 441)
    a. far more likely
    b. usually
    c. more likely
    d. less likely

TRUE-FALSE QUESTIONS

T F 1. Rates of illness and death are highest among the old and young. (p. 420)

T F 2. Class is generally more of a factor for health problems than race or ethnicity. (p. 421)

T F 3. Therapeutic use of drugs occurs when a person takes a drug for a specific purpose, such as reducing fever or a cough. (p. 421)

T F 4. Cocaine is the most extensively used illegal drug in the United States. (p. 422)

T F 5. HIV can be transmitted by an infected woman breast feeding her child. (p. 423)

T F 6. As a result of the Flexner Report, all African American medical schools in existence at the time of the report were awarded large federal grants in order to remain open. (p. 424)

T F 7. According to Paul Starr, third-party health insurance providers are the main reason for medical inflation. (p. 426)

T F 8. Managed care allows doctors more individual freedom in choosing the medical treatment for their patients. (p. 427)

T F 9. The Canadian health care system is an example of socialized medicine. (p. 428)

T F 10. The British health care system constitutes socialized medicine. (p. 429)

T F 11. Alternative medicine utilizes healing practices inconsistent with conventional medicine. (pp. 430-431)

T F 12. According to functionalist theorists, physicians hold a legal monopoly over medicine. (pp. 431-432)

T F 13. The subjective component of medicalization and demedicalization reflects the major concern of the symbolic interactionist perspective of health. (pp. 434-435)

T F 14. Acting to the functionalist perspectives, doctors have gained prestige through the classification of disease. (p. 436)

T F 15. Less than 15 percent of persons with a disability today were born with it. (p. 437)

SOCIOLOGY IN OUR TIMES: DIVERSITY ISSUES

1. Is health care a right or a privilege? Should the United States establish a national health care system for all Americans? Why? Why not?

2. Which health care system do you think is better for their citizens: Canada or Great Britain? Why?

3. How are ethnic/racial minority groups, women, and the aged affected by health care in the United States?

4. Explain this statement: "America has the best health care system that money can buy and therein lies the problem."

INTERNET EXERCISES

Use the search engine of your choice to assist you in completing the exercises listed below:

1. Compare the average life expectancy of a man and a woman in America versus a third-world country. Then discuss any possible reasons for why a gap between life expectancy may exist.

2. Similarly to exercise number one, compare the infant mortality rates of children in America versus those of a third-world country. Also, explain possible reasons for the statistics.

INFOTRAC COLLEGE EDITION ONLINE READINGS AND EXERCISES

Access the Infotrac College Edition web site. Once logged on type in "Drug that cuts off a tumor's blood supply" in the **keyword** search field. Once, the list of periodicals has appeared select the one by *Cancer Weekly*:

a. What does Avastin do to tumors?

b. What type of cancer is Avastin being test on and what phase of clinical testing is it at?

Return to the search field and enter "Healing Patients on a Spiritual, as well as Physical, Level" in the **keyword** search field. Hit enter and then select the article. Read the article and then answer the following questions:

a. What is an anacephalic fetus?

b. According to the article, are nurses and doctors allowed to become "spiritually" close to their patients? What is your opinion on the matter?

STUDENT CLASS PROJECTS AND ACTIVITIES

1. Collect at least 10 articles dealing with current issues of health and health care found in the text, such as alcoholism, concerns about diet and exercise, AIDS, current causes of death, the American health care system, the profession of medicine (including doctors, nurses, hospitals), health insurance, prepaid health care, and health care in other countries. The articles can come from

newspapers, professional journals, magazines, etc. You are to: (1) collect the articles; (2) write a summary explaining the message of each article; (3) write a person reaction to each article; (4) provide a conclusion and personal evaluation of the project, and (5) provide a bibliographic reference for each article selected. Submit the articles at the appropriate time.

2. Investigate a health care system found in another modern industrial country that has a reputation of having a fairly good, reliable health care delivery system. Some suggestions are: Great Britain, Canada, Germany, Sweden, Norway, and Japan. You are to describe in your paper: (1) the specific type(s) of health care system; (2) who receives the health care; (3) how the program is funded; (4) if personal choice is allowed--even for the higher income; (5) who the professionals are in the system; (6) their training; (7) how serious illness or disease is treated; (8) any other information pertinent to this research. In the writing of your paper, include a bibliography, a summary, a conclusion, and a personal evaluation of this project. Submit your paper to your instructor at the appropriate time.

3. The May/June 2004 issue of AARP, listed the 50 top hospitals in the United States, as cited by Consumer's Checkbooks, a nonprofit consumer education organization. This article summarizes the major findings of the publication, Consumer's Checkbook's Guide to Hospitals, which rates more than 4,500 hospitals nationwide. Read this article and summarize the major criteria and the major findings of the report. Are there any surprises? What constitutes the accreditation score, the physician's rating, and special offerings/programs/research of the top-ranked hospitals? List the top 25 hospitals in the nation. Next, discuss the top-ranked hospital, noting the explanations for its ranking. Write up your research following any specific directions provided by your instructor.

4. According to an article in Time magazine,(November 8, 1999) what were the ten causes of death and disabilities in the 1990's? What are the projections for the top ten causes of death and disability in the year 2020? Select one of these causes of the year 2020 and research that specific cause. Why is it a projected cause of death or disability? Why is it listed as a top ten projected cause of death or disability? What should be done to diminish the devastating effects of that cause? Submit your paper following any directions provided by your instructor.

ANSWERS TO PRACTICE TESTS FOR CHAPTER 14

Answers to Multiple Choice Questions

1. b The World Health Organization (WHO) defines health as a state of complete physical, mental, and social well-being. (pp 415-416)
2. a One aspect of institutional healing is health care and a health care delivery system in society. (p. 416)
3. c Health refers to the positive sense of complete well-being; while illness refers to an interference with health. (p. 416)
4. d Health care is any activity intended to improve health. (p. 416)
5. c Medicine is an institutionalized system for the scientific diagnosis, treatment, and prevention of illness. (p. 416)
6. c We do not hear about it as much now, but the epidemic continues. (p. 416)
7. d While life expectancy has increased, it is due to people living beyond childhood that it has increased. (pp. 416-417)
8. d Early deaths, including infant mortality are common. (p. 417)
9. b The World Health Organization reports that about two-thirds of the babies born during 2004 will die during the first month of life. (p. 418)
10. b The United States spends over $3,925 per person. (p. 419)
11. c There becomes a question as to where the money in the United States is spent regarding health care. Typically it is viewed as going toward bureaucracy. (p. 419)
12. a Social epidemiologists are people who study the causes and distribution of health, disease, and impairment in a population. (p. 420)
13. d Typically the target of social epidemiology (p.420)
14. b These are more common now than acute diseases. (p.420)
15. d Acute diseases are illnesses that strike suddenly and cause dramatic incapacitation, and sometimes death. (p. 420)
16. d All of the responses in "a" through "c" are some serious consequences of alcoholism. (p. 422)
17. b Nicotine is a toxic, dependency-producing drug that is more addictive than heroin. (p. 422)
18. c Gonorrhea and syphilis were the principal STDs in this country until the 1960s. (p. 423)
19. a More than two-thirds percent of all people with HIV/AIDS live in sub-Saharan Africa. (p. 424)
20. d The Flexner Report stated in 1910 that most existing medical schools were inadequate. (p. 424)
21. b Throughout most of the twentieth century, medical care in the United States was paid for on a fee-for-service basis. (p. 425)

22. c Medicaid is a jointly funded federal-state-local health care program for the poor. (p. 427)
23. a The only high-income nations without some form of universal health coverage for all its citizens are the United States and the Union of South Africa. (p. 426)
24. c An approach to health care that treats the whole person rather than just symptoms that occur is known as holistic medicine. (p. 430)
25. a Barefoot doctors are individuals who had little formal medical training in China and work in rural communities treating disease and illness. (p. 429)
26. b People in the sick role are expected to desire to get well. (p. 431)
27. d Highly questionable, however we do see the influence and power of medicine in the lives of Americans. (pp. 434-435)
28. c The results and effects of the labeling of behavior. (p. 435)
29. b There is more to it than simply this, but the field of medicine has lost some of its power due to other industries impacting their outcomes financially. (p. 435)
30. d Women with a disability are less likely to be covered by pension and health plans. (p. 441)

Answers to True-False Questions

1. True (p. 420)

2. True (p. 421)

3. True (p. 421)

4. False Marijuana is the most extensively used illegal drug in the United States. (p. 422)

5. True (p. 423)

6. False As a result of the Flexner Report all but two African-American medical schools then in existence were closed. (p. 424)

7. True (p. 426)

8. False Managed care requires doctors to get approval before they perform certain procedures. (p. 427)

9. False The Canadian health care system is not an example of socialized medicine; Canada has a universal health care system. (p. 428)

10. True (p. 429)

11. True (pp. 430-431)

12. False Conflict theorists focus on the legal monopoly that physicians hold over medicine as well as the medical-industrial complex in which they work. (pp. 431-432)

13. True (pp. 434-435)

14. False According to postmodernist approach, doctors have gained prestige through the classification of disease. (p. 436)

15. True (p. 437)

CHAPTER 15
POPULATION AND URBANIZATION

BRIEF CHAPTER OUTLINE

Demography: The Study of Population
- Fertility
- Mortality
- Migration
- Population Composition

Population Growth in Global Context
- The Malthusian Perspective
- The Marxist Perspective
- The Neo-Malthusian Perspective
- Demographic Transition Theory
- Other Perspectives on Population Change

A Brief Glimpse at International Migration Theories

Urbanization in Global Perspective
- Emergence and Evolution of the City
- Preindustrial Cities
- Industrial Cities
- Postindustrial Cities

Perspectives on Urbanization and the Growth of Cities
- Functionalist Perspectives: Ecological Models
- Conflict Perspectives: Political Economy Models
- Symbolic Interactionist Perspectives: The Experience of City Life

Problems in Global Cities

Urban Problems in the United States
- Divided Interests: Cities, Suburbs, and Beyond
- The Continual Fiscal Crisis of the Cities

Population and Urbanization in the Future

CHAPTER SUMMARY

Demography is the study of the size, composition, and distribution of the population. Population growth is the result of **fertility** (births), **mortality** (deaths), and **migration**. The **population composition** – the biological and social characteristics of a population is affected by changes in fertility, mortality, and migration. One measure of population composition is **sex ratio** – the number of males for every hundred females in a given population. For demographics, sex and age are significant characteristics; they are key indicators of fertility and mortality rates. Over two hundred years ago, Thomas Malthus warned that overpopulation would result in major global problems such as poverty and starvation. According the Marxist perspective, overpopulation occurs because of capitalists' demands for a surplus of workers to suppress wages and heighten workers' productivity. More recently, neo-Malthusians have re-emphasized the dangers of our population, depicting earth as a dying planet with too many people, too little food, compounded by environmental degradation. **Demographic transition** is the process by which some societies have moved from high birth and death rates to relatively low birth and death rates as a result of technological development. Other perspectives on population change include rational choice theory, the epidemiological transition, economic development, and the process of "westernization." In explaining international migration theories, the neoclassical economics approach, the new households economics of migration approach, conflict and world systems theory, all add to our knowledge of the way people migrate. *Urban sociology* is the study of social relationships and political and economic structures in the city. Cities are a relatively recent innovation when compared to the length of human existence. Because of their limited size, preindustrial cities tend to provide a sense of community and a feeling of belonging. The Industrial Revolution changed the size and nature of the city; people began to live close to the factories and to one another, which led to overcrowding and poor sanitation. In postindustrial cities, some people live and work in suburbs or outlying edge cities. Functionalist perspectives (ecological models) of urban growth include the *concentric zone model*, the *sector model*, and the *multiple-nuclei model*. According to the political economy models of conflict theorists, urban growth is influenced by capital investment decisions, power and resource inequality, class and class conflict, and government subsidy programs. Feminist theorists suggest that cities have *gender regimes*; women's lives are affected by both public and private patriarchy. Symbolic interactionists focus on the positive and negative aspects of peoples' experiences in the urban settings. Rapid population growth in many global cities is producing a wide variety of urban problems including overcrowding, environmental pollution and disappearance of farmland as well as creating a limit in the availability of basic public services. Urbanization, suburbanization, **gentrification**, and the growth of edge cities have had a dramatic impact on the U.S. population. Many central cities continue to experience fiscal crises that have resulted in cuts in services, lack of maintenance of the infrastructure, as well as loss of resources because of terrorist attacks.. Many cities and large urban areas have created a "disabling" environment for many people; access is critical in order for persons with a disability to become productive members of the community. Rapid global population growth is inevitable in the future. The urban population will triple as increasing numbers of people in lesser developed and developing nations migrate from rural areas to megacities that contain a high percentage of a region's population.

LEARNING OBJECTIVES

After reading Chapter 15, you should be able to:

1. Describe the study of demography and define the basic demographic concepts.

2. Explain the Malthusian perspective on population growth.

3. Discuss the Marxist perspective on population growth and compare it with the Malthusian perspective.

4. Describe the neo-Malthusian perspective on population growth and note the significance of zero population growth to this approach.

5. Discuss demographic transition theory and explain why this theory may not apply to population growth in all societies.

6. Trace the historical development of cities and identify the major characteristics of preindustrial, industrial, and postindustrial cities.

7. Discuss functionalist perspectives on urbanization and outline the major ecological models of urban growth.

8. Compare and contrast conflict and functionalist perspectives on urban growth.

9. Describe global patterns of urbanization in core, peripheral, and semiperipheral nations.

10. Explain symbolic interactionist perspectives on urban life and note the key assumptions of the major urban theorists.

11. Discuss the major problems facing urban areas in the United States today.

12. Explain the issues of the continual fiscal crises of the cities.

13. Define all the key terms in this chapter.

KEY TERMS (defined at page number shown and in glossary)

community 461
crude birth rate 448
crude death rate 449
democratic transition 456
demography 446
fecundity 447
fertility 447
gentrification 464
human ecology 463
invasion 463
metropolis 462
migration 450
mortality 449
population composition 451
population pyramid 452
sex ratio 451
succession 463
zero population growth 456

CHAPTER OUTLINE

I. DEMOGRAPHY: THE STUDY OF POPULATION
   A. **Demography** is a subfield of sociology that examines population size, composition, and distribution.
   B. **Fertility** is the actual level of childbearing for an individual or a population; **fecundity** is the potential number of children that could be born if every woman reproduced at her maximum biological capacity.
      1. The **crude birth rate** is the number of live births per 1,000 people in a population in a given year.
      2. In most areas of the world, women are having fewer children; women who have six or seven children tend to live in agricultural regions where children's labor is essential to the family's economic survival and child mortality rates are very high.
   C. A decline in **mortality** -- the incidence of death in a population -- has been the

primary cause of world population growth in recent years.

1. The **crude death rate** is the number of deaths per 1,000 people in a population in a given year.
2. The **infant mortality rate** is the number of deaths of infants under 1 year of age per 1,000 live births in a given year.

D. **Migration** is the movement of people from one geographic area to another for the purpose of changing residency.

1. While *immigration* is the movement of people into a geographic area to take up residency, *emigration* is the movement of people out of a geographic area to take up residency elsewhere.
2. *Push* and *pull* factors strongly influence migration patterns.
3. The **crude net migration rate** is the net number of migrants (total in-migrants minus total out-migrants) per 1,000 people in a population in a given year.

E. **Population composition** is the biological and social characteristics of a population, including age, sex, race, marital status, education, occupation, and income.

1. The **sex ratio** is the number of males for every hundred females in a given population; a sex ratio of 100 indicates an equal number of males and females.
2. A **population pyramid** is a graphic representation of the distribution of a population by sex and age.

II. POPULATION GROWTH IN GLOBAL CONTEXT

A. The Malthusian Perspective

1. According to Thomas Robert Malthus, the population (if left unchecked) would exceed the available food supply; population would increase in a geometric progression (2, 4, 8, 16 . . .) while the food supply would increase only by an arithmetic progression (1, 2, 3, 4 . . .).
2. This situation could end population growth and perhaps the entire population unless positive checks (such as famines, disease and wars) or preventive checks (such as sexual abstinence and postponement of marriage) intervened.

B. The Marxist Perspective

1. According to Karl Marx and Friedrich Engels, food supply does not have to be threatened by overpopulation; through technology, food for a growing population can be produced.
2. Overpopulation occurs because capitalists want a surplus of workers (an industrial reserve army) to suppress wages and force employees to be more productive.

C. The Neo-Malthusian Perspective

1. Neo-Malthusians (or "New Malthusians") reemphasized the dangers of overpopulation and suggested that an exponential growth pattern is occurring.
2. Overpopulation and rapid population growth result in global

environmental problems, and people should be encouraging **zero population growth** -- the point at which no population increase occurs from year to year because the number of births plus immigrants is equal to the number of deaths plus emigrants.

D. Demographic Transition Theory
   1. **Demographic transition** is the process by which some societies have moved from high birth and death rates to relatively low birth and death rates as a result of technological development.
   2. Demographic transition is linked to four stages of economic development:
      a. Stage 1: Preindustrial Societies -- little population growth occurs, high birth rates are offset by high death rates.
      b. Stage 2: Early Industrialization -- significant population growth occurs, birth rates are relatively high while death rates decline.
      c. Stage 3: Advanced Industrialization and Urbanization -- very little population growth occurs, both birth rates and death rates are low.
      d. Stage 4: Postindustrialization -- birth rates continue to decline as more women are employed full-time and raising children becomes more costly; population growth occurs slowly, if at all, due to a decrease in the birth rate and a stable death rate.
   3. Critics suggest that demographic transition theory may not accurately explain population growth in all societies; this theory may best explain growth in Western societies.

E. Other Perspectives on Population Change
   1. According to some scholars, the processes of industrialization and economic development are typically accompanied by *secularization*, is linked to a decline in fertility.
   2. *Rational choice theory*, which is based upon the assumption that people make decisions based upon a calculated cost-benefit analysis, suggests that as child rearing becomes more expensive, as societies modernize and urbanize, fertility rates decline.

III. A BRIEF GLIMPSE AT INTERNATIONAL MIGRATION THEORIES
   A. In explaining international migration theories, the neoclassical economics approach, the new households economics of migration approach, two conflict perspectives, and world systems theory, all add to our knowledge of the way people migrate
   B. These diverse theories of contemporary patterns of migration demonstrate that the reasons that people migrate are numerous and complex, involving processes that occur at the individual, family, and societal levels.

IV. URBANIZATION IN GLOBAL PERSPECTIVE
   A. *Urban sociology* is a subfield of sociology that examines social relationships and political and economic structures in the city.
   B. Emergence and Evolution of the City

1. Cities are a relatively recent innovation as compared with the length of human existence. According to Gideon Sjoberg, three preconditions must be present in order for a city to develop:
    a. A favorable physical environment
    b. An advanced technology that could produce a social surplus
    c. A well-developed social organization to provide social stability to the economic system.
2. Sjoberg places the first cities in the Mesopotamian region or areas immediately adjacent to it at about 3500 B.C.E; however, not all scholars agree on this point; some place the earliest city in Jericho.

C. Preindustrial Cities
1. The largest preindustrial city was Rome.
2. Preindustrial cities were limited in size because of crowded housing conditions, lack of adequate sewage facilities, limited food supplies, and lack of transportation to reach the city.
3. Many preindustrial cities had a sense of **community** -- a set of social relationships operating within given spatial boundaries that provide people with a sense of identity and a feeling of belonging, defined by Tonnies as *Gemeinschaft*.
4. Industrial cities were characterized by Tonnes as *Gesellschaft* – societies exhibiting impersonal and specialized relationships.

D. Industrial Cities
1. The nature of the city changed as factories arose and new forms of transportation and agricultural production made it easier to leave the countryside and move to the city.
2. New York City became the first U.S. **metropolis** -- one or more central cities and their surrounding suburbs that dominate the economic and cultural life of a region.
3. People lived in close proximity to factories so that they could walk to work; many lived in overcrowded conditions that lacked sanitation and a clean water supply.

E. Postindustrial Cities
1. Since the 1950s, postindustrial cities have emerged as the U.S. economy has gradually shifted from secondary (manufacturing) to tertiary (service and information processing) production.
2. Postindustrial cities are dominated by "light" industry, such as computer software manufacturing, information-processing services, educational complexes, medical centers, retail trade centers, and shopping malls.
3. On a global basis, cities such as New York, London, and Tokyo appear to fit the model of the postindustrial city; they have experienced a rapid growth in knowledge-based industries such as financial services.

V. PERSPECTIVES ON URBANIZATION AND THE GROWTH OF CITIES

A. Functionalist Perspectives: Ecological Models

1. Robert Park based his analysis of the city on **human ecology** -- the study of the relationship between people and their physical environment -- and found that economic competition produces certain regularities in land-use patterns and population distributions.
2. Concentric zone model
   a. Based on Park's ideas, Ernest W. Burgess developed a model that views the city as a series of circular zones, each characterized by a different type of land use, that developed from a central core: (1) the central business district and cultural center; (2) the zone of transition -- houses where wealthy families previously lived that have now been subdivided and rented to persons with low incomes; (3) working-class residences and shops, and ethnic enclaves; (4) homes for affluent families, single-family residences of white-collar workers, and shopping centers; and (5) a ring of small cities and towns comprised of estates owned by the wealthy and houses of commuters who work in the city.
   b. Two important ecological processes occur: **invasion** is the process by which a new category of people or type of land use arrives in an area previously occupied by another group or land use, and **succession** is the process by which a new category of people or type of land use gradually predominates in an area formerly dominated by another group or activity.
   c. **Gentrification** is the process by which members of the middle and upper-middle classes, especially whites, move into the central city area and renovate existing properties.
3. Sector Model
   a. Homer Hoyt's *sector model* emphasizes the significance of terrain and the importance of transportation routes in the layout of cities.
   b. Residences of a particular type and value tend to grow outward from the center of the city in wedge-shaped sectors with the more expensive residential neighborhoods located along the higher ground near lakes and rivers, or along certain streets that stretch from the downtown area.
   c. Industrial areas are located along river valleys and railroad lines; middle class residences exist on either side of wealthier neighborhoods; lower class residential areas border the central business area and the industrial areas.
4. Multiple-Nuclei Model
   a. According to Chauncey Harris and Edward Ullman, cities have numerous centers of development; as cities grow, they annex outlying townships.
   b. In addition to the central business district, other nuclei develop around entities such as an educational institution or a medical complex; residential neighborhoods may exist close to or far away from these nuclei.

5. Contemporary Urban Ecology.
   a. Amos Hawley viewed urban areas as complex social systems in which growth patterns are based on advances in transportation and communication.
   b. *Social area analysis* examines urban populations in terms of economic status, family status, and ethnic classification.

B. Conflict Perspectives: Political Economy Models
1. According to Marx, cities are arenas in which the intertwined processes of class conflict and capital accumulation take place; class consciousness is more likely to occur in cities where workers are concentrated.
2. Three major themes are found in political economy models:
   a. Patterns of urban growth and decline are affected by: (1) economic factors such as capitalist investments; and (2) political factors, including governmental protection of private property and promotion of the interests of business elites and large corporations.
   b. Urban space has both an exchange value and a use value: (1) exchange value refers to the profits industrialists, developers, and bankers make from buying, selling, and developing land and buildings; (2) use value is the utility of space, land, and buildings for family life and neighborhood life.
   c. Structure and agency are both important in understanding how urban development takes place: (1) structure refers to institutions such as state bureaucracies and capital investment circuits that are involved in the urban development process; and (2) agency refers to human actors who participate in land use decisions, including developers, business elites, and activists protesting development.
3. According to political economy models, urban growth is influenced by capital investment decisions, power and resource inequality, class and class conflict, and government subsidy programs.
4. Gender Regimes in Cities
   a. According to feminist perspectives, urbanization reflects the workings of the political economy and patriarchy.
   b. Different cities have different *gender regimes* -- prevailing ideologies of how women and men should think, feel, and act; how access to positions and control of resources should be managed; and how women and men should relate to each other.
   c. Gender intersects with class and race as a form of oppression, especially for lower-income women of color who live in central cities.

C. Symbolic Interactionist Perspectives: The Experience of City Life
1. Simmel's View of City Life
   a. According to Georg Simmel, urban life is highly stimulating; it shapes people's thoughts and actions.
   b. However, many urban residents avoid emotional involvement with each other and try to ignore events taking place around them.

c. City life is not completely negative; urban living can be liberating -- people have opportunities for individualism and autonomy.

2. Urbanism as a Way of Life
   a. Louis Wirth suggested that urbanism is a "way of life." *Urbanism* refers to the distinctive social and psychological patterns of city life.
   b. Size, density, and heterogeneity result in an elaborate division of labor and in spatial segregation of people by race/ethnicity, class, religion, and/or lifestyle; a sense of community is replaced by the "mass society" -- a large-scale, highly institutionalized society in which individuality is supplanted by mass media, faceless bureaucrats, and corporate interests.
3. Gans's Urban Villagers
   a. According to Herbert Gans, not everyone experiences the city in the same way; some people develop strong loyalties and a sense of community within central city areas that outsiders may view negatively.
   b. Five major categories of urban dwellers are: (1) *cosmopolites* -- students, artists, writers, musicians, entertainers, and professionals who choose to live in the city because they want to be close to its cultural facilities; (2) *unmarried people and childless couples* who live in the city because they want to be close to work and entertainment; (3) *ethnic villagers* who live in ethnically segregated neighborhoods; (4) *the deprived* -- individuals who are very poor and see few future prospects; and (5) *the trapped* -- those who cannot escape the city, including downwardly mobile persons, older persons, and persons with addictions.
4. Gender and City Life
   a. According to Elizabeth Wilson, some men view the city as sexual space in which women, based on their sexual desirability and accessibility, are categorized as prostitutes, lesbians, temptresses, or virtuous women in need of protection.
   b. More affluent, dominant group women are more likely to be viewed as virtuous women in need of protection while others are placed in less desirable categories.
5. Cities and Persons with a Disability
   1. Many cities and urban areas create a “disabling” environment for many people.
   2. Harlan Hahn suggests that historical patterns in the dynamics of capitalism contributed to discrimination against persons with disabilities, and this legacy remains today.

VI. PROBLEMS IN GLOBAL CITIES

A. Natural increases in population account for two-thirds of new urban growth, and rural-to-urban migration accounts for the remainder.

B. Rapid global population growth in Latin American and other regions is producing a wide variety of urban problems, including overcrowding, environmental pollution and the disappearance of farmland.
C. Most African countries and many countries in South America and the Caribbean are peripheral nations.
D. Cities in semipheripheral nations—such as India, Iran, and Mexico are confronted with unprecedented population growth.

VII. URBAN PROBLEMS IN THE UNITED STATES

A. Divided Interests: Cities, Suburbs, and Beyond
   1. Since World War II, the U.S. population has shifted dramatically as many people have moved to the suburbs.
   2. Suburbanites rely on urban centers for employment and some services but pay property taxes to suburban governments and school districts; some affluent suburbs have state-of-the-art school districts and infrastructure while central city services and school districts lack funds.
   3. Race, Class, and Suburbs
      a. The intertwining impact of race and class is visible in the division between central cities and suburbs.
      b. Most suburbs are predominantly white; many upper-middle and upper-class suburbs remain virtually all white; people of color who live in suburbs often are resegregated.
   4. Beyond the Suburbs
      a. *Edge cities* initially develop as residential areas beyond central cities and suburbs; then retail establishments and office parks move into the area and create an unincorporated edge city.
      b. Corporations move to edge cities because of cheaper land and lower utility rates and property taxes.
   5. Likewise, Sunbelt cities grew in the 1970s, as millions moved from the north and northeastern states to southern and western states where there were more jobs and higher wages, lower taxes, pork-barrel programs funded by federal money that created jobs and encouraged industry, and the presence of high technology industries.

B. The Continual Fiscal Crisis of the Cities
   1. The largest cities in the United States have faced periodic fiscal crisis for many years. In the twenty-first century, these crises have been intensified by higher employee health care and pension costs, declining revenue, and increased expenditures for public safety and homeland security.
   2. Many cities have cut back on spending in areas other than public safety.
   3. Even if the U.S. economy improves significantly in the near future, analysts believe that the positive effects of such a rebound will not improve the budgetary problems of our cities and towns for a number of years.

VIII. POPULATION AND URBANIZATION IN THE FUTURE

A. Rapid global population growth is inevitable: although death rates have declined in many low-income nations, birth rates have not had a corresponding decline.

B. In the future, low-income countries will have an increasing number of poor people. The world's population will double, the urban population will triple as people migrate from rural to urban areas.

C. At the macrolevel, we can do little about population and urbanization; at the microlevel, we may be able to exercise some degree of control over our communities and our own lives.

ANALYZING AND UNDERSTANDING THE BOXES

After reading the chapter and studying the outline, re-read the boxes and write down key points and possible questions for class discussion.

Sociology and Everyday Life: How Much Do You Know About U.S. Immigration?

Key Points:

Discussion Questions:

1.

2.

3.

Framing Immigration in the Media: Media Framing and Public Opinion

Key Points:

Discussion Questions:

1.

2.

3.

You Can Make a Difference: Working Toward Better Communities

Key Points:

Discussion Questions:

1.

2.

3.

PRACTICE TESTS

MULTIPLE CHOICE QUESTIONS

Select the response that best answers the question or completes the statement:

1. The world's population reached __________ people in 2006 (p. 446).
   a. 6.5 million
   b. 60.5 million
   c. 6.5 billion
   d. 60.5 billion

2. Demography is a subfield of sociology that examines: (p. 446)
   a. population size.
   b. population composition.
   c. population distribution.
   d. all of the above.

3. The world's population is increasing by approximately _____ people per year. (p. 446)
   a. 7.6 million
   b. 76 million
   c. 760 million
   d. 7 billion

4. ______ is the actual level of childbearing for an individual or a population, while _____ is the potential number of children that could be born if every woman reproduced at her maximum biological capacity. (p. 447)
   a. Birth rate fertility
   b. Fertility rate birth rate
   c. Fertility fecundity
   d. Fecundity fertility

5. Changes in population occur as a result of __________. (p. 447)
   a. fertility
   b. mortality
   c. migration
   d. (all of the above)

6. The primary biological factor affecting ________ is the number of women of childbearing age.
   a. fertility
   b. mortality
   c, migration
   d. demography

7. Sociologists usually consider __________ to be the childbearing age of women. (p. 447)
   a. 11-35
   b. 15-45
   c. 18-40
   d. 18-45

8. Social factors influencing __________ include the roles available to women in a society and prevalent viewpoints regarding what constitutes the "ideal" family size. (p. 447)
   a. fertility
   b. mortality
   c. migration
   d. demography

9. Based on biological capacity alone, most women could produce __________ children during their childbearing years. (p. 447)
   a. about 6
   b. about 12
   c. about 15
   d. 20 or more

10. The primary cause of world population growth in recent years is a/an: (p. 449)
    a. increase in the birth rate.
    b. decline in the mortality.
    c. decline in all infectious diseases.
    d. increase in post baby boom birth rates.

11. Many serious diseases have been virtually eliminated in high-income, developed nations as a result of __________. (p. 449)
    a. improved sanitation
    b. improved personal hygiene
    c. vaccinations
    d. (all of the above)

12. Which of the following factors tend to limit fertility? (p. 447)
    a. infanticide
    b. contraception
    c. sterilization
    d. (all of the above)

13. The basic measure of fertility is the __________, which is the number of live births per 1,000 people in a population in a given year. (p. 448)
    a. crude birth rate
    b. corrected birth rate
    c. population rate
    d. demographic rate

14. The average lifetime in years of people born in a specific year is known as: (p. 450)
    a. life expectancy.
    b. the crude mortality rate.
    c. the longevity table.
    d. ages-specific death rates.

15. ______ is the movement of people out of a geographic area to take up residency elsewhere. (p. 450)
    a. Immigration
    b. Emigration
    c. Transmigration
    d. Ex-migration

16. According to the perspective on population by Thomas Malthus,: (p. 452)
    a. the population would increase in a geometric progression while the food supply would increase in an arithmetic progression.
    b. the population would increase in an arithmetic progression while the food supply would increase in a geometric progression.
    c. the food supply is not threatened by overpopulation because technology makes it possible to produce the food and other goods needed to meet the demands of a growing population.
    d. societies move through a process of demographic transition.

17. According to Karl Marx and Friedrich Engels' perspective on population: (pp. 452-453)
    a. the population would increase in a geometric progression while the food supply would increase in an arithmetic progression.
    b. the population would increase in an arithmetic progression while the food supply would increase in a geometric progression.
    c. the food supply is not threatened by overpopulation because technology makes it possible to produce the food and other goods needed to meet the demands of a growing population.
    d. societies move through a process of demographic transition.

18. According to the demographic transition theory, significant population growth occurs because birth rates are relatively high while death rates decline in the______stage of economic development. (p. 456)
    a. preindustrial
    b. early industrial
    c. advanced industrial
    d. postindustrial

19. All of the following statements regarding preindustrial cities are true, *except*: (p. 461)
    a. the largest preindustrial city was Rome.
    b. preindustrial cities were limited in size because of crowded housing conditions and a lack of adequate sewage facilities.
    c. food supplies were limited in preindustrial cities.
    d. preindustrial cities lacked a sense of community.

20. The *Gemeinschaft* and *Gesellschaft* typology originated with: (p. 461)
    a. Emile Durkheim.
    b. Gideon Sjoberg.
    c. Max Weber.
    d. Ferdinand Tonnies.

21. ______refers to one or more central cities and their surrounding suburbs that dominate the economic and cultural life of a region. (p. 462)
    a. Urban sprawl
    b. Megalopolis
    c. Metropolis
    d. Urbanization

22. Postindustrial cities are characterized by: (p. 462)
    a. "light" industry, information-processing services, educational complexes, retail trade centers, and shopping malls.
    b. the growth of the factory system.
    c. agricultural production.
    d. "heavy" industry, such as automobile manufacturing.

23. Ecological models of urban growth are based on a _______ perspective. (p. 463)
    a. functionalist
    b. conflict
    c. neoMarxist
    d. symbolic interactionist

24. An upper-middle class doctor who moves her family from the suburbs into the central city to renovate an older home is an example of: (p. 464)
    a. succession.
    b. gentrification.
    c. exsuburbanization.
    d. downward immigration.

25. All of the following are ecological models of urban growth, *except* the: (pp. 463-465)
    a. concentric zone model.
    b. sector model.
    c. political economy model.
    d. multiple nuclei model.

26. According to political economy models, urban growth is: (pp. 465-466)
    a. influenced by terrain and transportation.
    b. based on the clustering of people who share similar characteristics.
    c. linked with peaks and valleys in the economic cycle.
    d. influenced by capital investment decisions, power and resource inequality, and government subsidy programs.

27. _______ refers to the tendency of some neighborhoods, cities, or regions to grow and prosper while others stagnate and decline. (p. 466)
    a. Invasion
    b. Succession
    c. Gentrification
    d. Uneven development

28. Sociologist ____ has argued that urbanism is "a way of life." (p. 467)
    a. Georg Simmel
    b. Herbert Gans
    c. Louis Wirth
    d. Mark Gottdiener

29. According to sociologists, edge cities: (p. 473)
    a. initially develop as industrial parks.
    b. drain taxes from central cities and older suburbs.
    c. have existed since World War II.
    d. always correspond to municipal boundaries.

30. A 2004 survey of 328 cities found that _____ of those cities were less able to meet their financial needs than in the previous year. (p. 473)
    a. 50 percent
    b. 60 percent
    c. 70 percent
    d. 80 percent

TRUE-FALSE QUESTIONS

T F 1. The world's population is increasing by more than 76 million people per year. (p. 446)

T F 2. In most areas of the world, women are having more children. (p. 449)

T F. 3. The primary cause of world population growth in recent years has been a decline in mortality. (p. 449)

T F 4. The top leading cause of death in the United States is HIV. (p. 450)

T F 5. Pull factors of migration include political unrest and war. (p. 451)

T F 6. Sex ratio is the number of females per 100 males in a given population. (p. 451)

T F 7. Population pyramids are graphic representations of the distribution of a population by sex and race. (p. 452)

T F 8. According to the Marxist perspective, overpopulation occurs because capitalists desire to have a surplus of workers so as to suppress wages and increase workers' productivity. (p. 456)

T F 9. According to Malthus, famine is an example of a positive check. (p. 452)

T F 10. The sector model views the city as a series of circular areas or zones, each characterized by a different type of land use that developed from a central core. (p. 464)

T F 11. According to conflict theorists, cities grow and decline by chance. (p. 465)

T F 12. According to feminist theorists, public patriarchy may be perpetuated by cities through policies that limit women's access to paid work and public transportation. (pp. 466-467)

T F 13. Herbert Gans has suggested that almost all city dwellers live in urban areas by choice. (p. 468)

T F 14. Nationally, most suburbs are predominantly white. (p. 472)

T F 15. Redlining makes the acquiring of certain properties virtually impossible. (p. 472)

SOCIOLOGY IN OUR TIMES: DIVERSITY ISSUES

1. How can the study of population and urbanization better prepare us for the future? Do you think suburbs will continue to grow, or will people begin to return to the inner city? What do you think will be the world's most populated city in 2025?

2. Is the community or city in which you live the "same" for men and women? Are women and men equally "safe" on the streets? Where do you feel most secure? Least secure? Does gender interact with class and race as a form of oppression for lower-income women of color?

3. How are race and class intertwined with the contemporary problems of cities and suburbs? Are people in your community or city segregated, either voluntarily or involuntarily, by race and class?

INTERNET EXERCISES

Log on to a search engine and complete the following statistics about fertility rates:

1. What country in the world has the highest fertility rate?

2. What country has the lowest fertility rate?

3. What could be a reason for the high and low rates?

INFOTRAC COLLEGE EDITION ONLINE READINGS AND EXERCISES

Log onto the Infotrac College Edition website to answer the following questions. Once logged on, type " UN Population Division and Replacement Migration " into the **keyword** search field. Once the list appears, click to read the article, then answer the following questions

a. What is the most salient demographic projection prepared by the Population Division of the United Nations?

b. What countries are expected to decrease in population size?

c. Why are migrants needed to maintain the populations of specific countries?

Now return to the **keyword** search field and type “The Great Cities of the Future ". Now scroll through the articles and choose the one by Mic Kinley Conway. Read the article and answer the following questions:

a. What is a "super-city"?

b. What are the vital elements for super-cities? List eighteen such cities that might qualify by the year 2015.

STUDENT CLASS PROJECTS AND ACTIVITIES

1. Utilizing the most current U.S. Census data, construct a population pyramid of the United States, your home state, and the city in which you live or reside, or of any city in the United States. Provide the pyramids, interpretations of the statistics, a summary of the data, a conclusion from each pyramid constructed, and bibliographical references. Submit your project to your instructor at the appropriate time.

2. Construct a demographic analysis of the state in which your college or university is located and compare this with an analysis of the entire United States. Present data on the size of the population, the numbers and percentage of increase or decrease over the past twenty years, the birth rate, the death rate, sex ratio, infant mortality rate, cause of death, and life expectancy. Provide a summary of the data, a conclusion, and an evaluation of this project. Submit your project at the appropriate time to your instructor.

ANSWERS TO PRACTICE TESTS FOR CHAPTER 15

Answers to Multiple Choice Questions

1. c Most of the growth is occurring in developing nations. (p. 446)
2. d Demography is a subfield of sociology that examines population size, population composition, and population distribution. Thus "all of the above" is the best answer. (p. 446)
3. b Due in large part to the reduction in infant mortality and improvements in sanitation. (p. 446)
4. c Fertility is the actual level of childbearing for an individual or a population, while fecundity is the potential number of children that could be born if every woman reproduced at her maximum biological capacity. (p. 447)
5. d All of the above apply. (p. 447)
6. a Which is usually between 15 and 45. (p. 447)
7. b This is the most common age bearing group. (p. 447)
8. a The social always impacts the physical. (p. 447)
9. d Given that childbearing years cover a 30 year span, this is not an unusual number. (p. 447)
10. b The primary cause of world population growth in recent years is a decline in the mortality. (p. 449)
11. d All of the above apply. (p. 449)
12 d All of the above apply. (p. 447)
13. a This is typically used in demographic research. (p.448)
14. a The average lifetime in years of people born in a specific year is known as life expectancy. (p. 450)
15. b Emigration is the movement of people out of a geographic area to take up residency elsewhere. (p. 450)

16. a According to the perspective on population of Thomas Malthus, , the population would increase in a geometric progression while the food supply would increase in an arithmetic progression. (p. 452)
17. c According to Karl Marx and Friedrich Engel's perspective on population, the food supply is not threatened by overpopulation because technology makes it possible to produce the food and other goods needed to meet the demands of a growing population. (p. 452-453)
18. b According to the demographic transition theory, significant population growth occurs because birth rates are relatively high while death rates decline in the early industrial stage of economic development. (pp. 456)
19. d All of the following statements regarding preindustrial cities are true, *except* preindustrial cities lacked a sense of community. (p. 461)
20. d The *Gemeinschaft* and *Gesellschaft* typology originated with Ferdinand Tonnies. (p. 461)
21. c Metropolis refers to one or more central cities and their surrounding suburbs that dominate the economic and cultural life of a region. (p. 462)
22. a Postindustrial cities are characterized by "light" industry, information-processing services, educational complexes, retail trade centers, and shopping malls. (p. 462)
23. a Ecological models of urban growth are based on a functionalist perspective. (p. 463)
24. b An upper middle-class doctor who moves from the suburbs into the central city to renovate an older home is an example of gentrification. (p. 464)
25. c All of the following are ecological models, *except* the political economy model. (pp. 463-465)
26. d According to political economy models, urban growth is influenced by capital investment decisions, power and resource inequality, and government subsidy programs. (pp. 465-466)
27. d Uneven development refers to the tendency of some neighborhoods, cities, or regions to grow and prosper while others stagnate and decline. (p. 466)
28. c Sociologist Louis Wirth has argued that urbanism is a "way of life." (p. 467)
29. b According to sociologists, edge cities drain taxes from central cities and older suburbs. (p. 473)
30. d The 2004 survey of 328 cities found that 80 percent of those cities were less able to meet their financial needs than in the previous year. (p. 473)

Answers to True-False Questions

1. True (p. 446)

2. False In most areas of the world, women are having fewer children. (p. 449)

3. True (p. 449)

4. False Heart disease is the leading cause of death in the United States. (p. 450)

5. False Push factors of migration include political unrest and war. (p. 451)

6. False Sex ratio is the number of *males* for every 100 females in a given population. (p. 451)

7. False Population pyramids are graphic representations of the distribution of a population by *sex* and *age.* (p. 452)

8. True (p. 456)

9. True (p. 452)

10. False The concentric zone model views the city as a series of circular areas or zones, each characterized by a different type of land use, that developed from a central core. (p. 464)

11. False According to conflict theorists, cities do not grow and decline by chance. Rather, they are the product of specific decisions made by members of the capitalist class and political elites. (p. 465)

12. True (pp. 466-467)

13. False Although Herbert Gans has suggested that some city dwellers (especially cosmopolites and unmarried people and childless couples who want to live close to work and entertainment) do live in the cities by choice, others (including the deprived and the trapped) can find no escape from the city. (p. 468)

14. True (p. 472)

15. True (p. 472)

CHAPTER 16
COLLECTIVE BEHAVIOR, SOCIAL MOVEMENTS AND SOCIAL CHANGE

BRIEF CHAPTER OUTLINE

Collective Behavior
- Conditions for Collective Behavior
- Dynamics of Collective Behavior
- Distinctions Regarding Collective Behavior
- Types of Crowd Behavior
- Explanations of Crowd Behavior
- Mass Behavior

Social Movements
- Types of Social Movements
- Stages in Social Movements

Social Movement Theories
- Relative Deprivation Theory
- Value-Added Theory
- Resource Mobilization Theory
- Social Constructionist Theory: Frame Analysis
- New Social Movement Theory

Social Change In The Future
- The Physical Environment and Change
- Population and Change
- Technology and Change
- Social Institutions and Change
- A Few Final Thoughts

CHAPTER SUMMARY

Social change is the alteration, modification, or transformation of public policy, culture, or social institutions over time. Such change usually is brought about by **collective behavior** -- voluntary, often spontaneous activity that is engaged in by a large number of people and typically violates dominant-group norms and values. A **crowd** is a relatively large number of people who are in one another's immediate vicinity. Five categories of crowds have been identified: (1) *casual crowds* are relatively large gatherings of people who happen to be in the same place at the same time; (2) *conventional crowds* are comprised of people who specifically come together for a scheduled event and thus share a common focus; (3) *expressive crowds* provide opportunities for the expression of some strong emotion; (4) *acting crowds* are collectivities so intensely focused on a specific purpose or object that they may erupt into violent or destructive behavior; and (5) *protest crowds* are gatherings of people who engage in activities intended to achieve specific political goals. Protest crowds sometime participate in **civil disobedience** -- nonviolent action that seeks to change a policy or law by refusing to comply with it. Explanations of crowd behavior include contagion theory, social unrest and circular reaction, convergence theory, and emergent norm theory. Examples of **mass behavior** -- collective behavior that takes place when people respond to the same event in much the same way -- include rumors, gossip, mass hysteria, fads, fashions, and public opinion. The major types of **social movements** -- organized groups that act consciously to promote or resist change through collective action -- are reform movements, revolutionary movements, religious movements, alternative movements, and resistance movements. Sociological theories explaining social movements include relative deprivation theory, value-added theory, resource mobilization theory, and recent emerging perspectives. Social change produces many challenges that remain to be resolved: environmental problems, changes in the demographics of the population, and new technology that benefits some -- but not all -- people. As we head into the future, we must use our sociological imaginations to help resolve these problems, not only in our country, but worldwide.

LEARNING OBJECTIVES

After reading Chapter 16, you should be able to:

1. Define collective behavior and describe the conditions necessary for such behavior to occur.

2. Distinguish between crowds and masses and identify casual, conventional, expressive, acting, and protest crowds.

3. Discuss the key elements of these four explanations of collective behavior: contagion theory, social unrest and circular reaction, convergence theory, and emergent norm theory.

4. Define mass behavior and describe the most frequent types of this behavior.

5. Describe social movements and note when and where they are most likely to develop.

6. Differentiate among the five major types of social movements based on their goals and the amount of change they seek to produce.

7. Identify the stages in social movements and explain why social movements may be an important source of social change.

8. Compare relative deprivation theory and value-added theory as explanations of why people join social movements.

9. State the key assumptions of resource mobilization theory.

10. Explain the basis and the assumptions of social constructionist theory and explain how problems are framed from this perspective.

11. Describe the "new social movement" theory and give examples of issues of concern to participants in these movements.

12. Describe the effects of physical environment, population trends, technological development, and social institutions on social change in the future.

KEY TERMS (defined at page number shown and in glossary)

civil disobedience 485
collective behavior 480
crowd 483
environmental racism 498
gossip 489
mass 483
mass behavior 487
mob 485
panic 485
propaganda 491
public opinion 490
riot 485
rumors 488
social change 480
social movement 492

CHAPTER OUTLINE

I. COLLECTIVE BEHAVIOR
   A. **Social change** is the alteration, modification, or transformation of public policy, culture, or social institutions over time; such change usually is brought about by **collective behavior** -- relatively spontaneous, unstructured activity that typically violates established social norms.
   B. Conditions for Collective Behavior
      1. Collective behavior occurs as a result of some common influence or stimuli which produces a response from a collectivity -- a relatively large number of people who mutually transcend, bypass, or subvert established institutional patterns and structures.
      2. Major factors that contribute to the likelihood that collective behavior will occur are:
         a. Structural factors that increase the chances of people responding in a particular way;
         b. Timing;
         c. A breakdown in social control mechanisms and a corresponding feeling of normlessness;
   C. Dynamics of Collective Behavior
      1. People may engage in collective behavior when they find that their problems are not being solved through official channels; as the problem appears to grow worse, organizational responses become more defensive and obscure.
      2. People's attitudes are not always reflected in their political and social behavior.
      3. People act collectively in ways they would not act singly due to:
         a. The noise and activity around them.
         b. A belief that it is the only way to fight those with greater power and resources.
   D. Distinctions Regarding Collective Behavior
      1. People engaging in collective behavior may be a:
         a. **Crowd** -- a relatively large number of people who are in one another's immediate face-to-face presence; or

b. **Mass** -- a number of people who share an interest in a specific idea or issue but who are not in one another's immediate physical vicinity.

2. Collective behavior also may be distinguished by the dominant emotion expressed (e.g., fear, hostility, joy, grief, disgust, surprise, or shame).

E. Types of Crowd Behavior

1. Herbert Blumer divided crowds into four categories:

a. *Casual crowds* -- relatively large gatherings of people who happen to be in the same place at the same time; if they interact at all, it is only briefly.

b. *Conventional crowds* -- people who specifically come together for a scheduled event and thus share a common focus.

c. *Expressive crowds* -- people releasing their pent-up emotions in conjunction with others who experience similar emotions.

d. *Acting crowds* -- collectivities so intensely focused on a specific purpose or object that they may erupt into violent or destructive behavior. Examples:

(1) A **mob** -- a highly emotional crowd whose members engage in, or are ready to engage in, violence against a specific target which may be a person, a category of people, or physical property.

(2) A **riot** -- violent crowd behavior fueled by deep-seated emotions but not directed at a specific target.

(3) A **panic** -- a form of crowd behavior that occurs when a large number of people react with strong emotions and self-destructive behavior to a real or perceived threat.

2. To these four types of crowds, Clark McPhail and Ronald T. Wohlstein added *protest crowds* -- crowds that engage in activities intended to achieve specific political goals.

a. *Protest crowds* sometimes take the form of **civil disobedience** -- nonviolent action that seeks to change a policy or law by refusing to comply with it.

b. At the grassroots level, protests often are seen as the only way to call attention to problems or demand social change.

F. Explanations of Crowd Behavior

1. According to *contagion theory*, people are more likely to engage in antisocial behavior in a crowd because they are anonymous and feel invulnerable; Gustave Le Bon argued that feelings of fear and hate are *contagious* in crowds because people experience a decline in personal responsibility.

2. According to Robert Park, social unrest is transmitted by a process of *circular reaction* -- the interactive communication between persons in such a way that the discontent of one person is communicated to another who, in turn, reflects the discontent back to the first person.

3. *Convergence* theory focuses on the shared emotions, goals, and beliefs many people bring to crowd behavior.

a. From this perspective, people with similar attributes find a collectivity of like-minded persons with whom they can release their underlying personal tendencies.

b. Although people may reveal their "true selves" in crowds, their behavior is not irrational; it is highly predictable to those who share similar emotions or beliefs.

4. According to Ralph Turner's and Lewis Killian's *emergent norm theory*, crowds develop their own definition of the situation and establish norms for behavior that fits the occasion.

a. Emergent norms occur when people define a new situation as highly unusual or see a long-standing situation in a new light.

b. Emergent norm theory points out that crowds are not irrational; new norms are developed in a rational way to fit the needs of the immediate situation.

G. Mass Behavior

1. **Mass behavior** is collective behavior that takes place when people (who often are geographically separated from one anther) respond to the same event in much the same way. The most frequent types of mass behavior are rumors, gossip, mass hysteria, public opinion, fashions, and fads.

2. **Rumors** are unsubstantiated reports on an issue or subject while **gossip** consists of rumors about the personal lives of individuals.

3. *Mass hysteria* is a form of dispersed collective behavior that occurs when a large number of people react with strong emotions and self-destructive behavior to a real or perceived threat; many sociologists believe this behavior is best described as a panic with a dispersed audience.

4. *Fads* and *Fashions*.

a. A *fad* is a temporary but widely copied activity enthusiastically followed by large numbers of people.

b. *Fashion* is a currently valued style of behavior, thinking, or appearance. Fashion also applies to art, music, drama, literature, architecture, interior design, and automobiles, among other things.

5. **Public opinion** consists of the attitudes and beliefs communicated by ordinary citizens to decision makers (as measured through polls and surveys based on interviews and questionnaires).

a. Even on a single topic, public opinion will vary widely based on characteristics such as race/ethnicity, religion, region of the country, social class, education level, gender, and age.

b. As the masses attempt to influence elites and visa versa, a two-way process occurs with the dissemination of **propaganda** -- information provided by individuals or groups that have a vested interest in furthering their own cause or damaging an opposing one.

II. SOCIAL MOVEMENTS

A. A **social movement** is an organized group that acts consciously to promote or resist change through collective action.

B. Types of Social Movements

1. *Reform* movements seek to improve society by changing some specific

aspect of the social structure.

2. *Revolutionary* movements seek to bring about a total change in society.
3. *Religious* movements seek to produce radical change in individuals and typically are based on spiritual or supernatural belief systems.
4. *Alternative* movements seek limited change in some aspect of people's behavior (e.g., a movement that attempts to get people to abstain from drinking alcoholic beverages).
5. *Resistance* movements seek to prevent or to undo change that already has occurred.

C. Stages in Social Movements

1. In the *preliminary* stage, widespread unrest is present as people begin to become aware of a threatening problem. Leaders emerge to agitate others into taking action.
2. In the *coalescence* stage, people begin to organize and start making the threat known to the public. Some movements become formally organized at local and regional levels.
3. In the *institutionalization* stage, an organizational structure develops, and a paid staff (rather than volunteers) begins to lead the group.

III. SOCIAL MOVEMENT THEORY

A. *Relative deprivation* theory asserts that people who suffer relative deprivation are likely to feel that a change is necessary and to join a social movement in order to bring about that change.

B. According to Neal Smelser's *value-added theory*, six conditions are necessary and sufficient to produce social movements when they combine or interact in a particular situation:

1. Structural conduciveness
2. Structural strain
3. Spread of a generalized belief
4. Precipitating factors
5. Mobilization for action
6. Social control factors

C. *Resource mobilization* theory focuses on the ability of a social movement to acquire resources (money, time and skills, access to the media, etc.) and mobilize people to advance the cause.

D. Social Constructionist theory is based on the assumption that social movements are interactive, symbolically defined, and a negotiated process.

1. This process involves participants, opponents, and bystanders.
2. Research based on this perspective often investigates how problems are framed and what names they are given.
3. Four distinct frame alignment processes occur in social movements:
   a. frame bridging;
   b. frame amplification;
   c. frame extension; and
   d. frame transformation

E. New social movement theory looks at a diverse array of collective actions and the manner in which those actions are based in politics, ideology, and culture.

1. It incorporates sources of identity, including race, class, gender, and sexuality, as sources of collective action and social movements.
2. Examples of already existing "new social movements" include ecofeminism and environmental justice movements.
   a. According to ecofeminists, patriarchy is a root cause of environmental problems because it contributes to a belief that nature is to be possessed and dominated, rather than treated as a partner.
   b. Environmental justice movements focus on the issue of **environmental racism** -- the belief that a disproportionate number of hazardous facilities (including industries such as waste disposal/treatment and chemical plants) are placed in low-income areas populated primarily by people of color.

IV. SOCIAL CHANGE IN THE FUTURE

A. The Physical Environment and Change: Changes in the physical environment often produce changes in the lives of people; in turn, people can make dramatic changes in the physical environment, over which we have only limited control.

B. Population and Change: Changes in population size, distribution, and composition affect the culture and social structure of a society and change the relationships among nations.

C. Technology and Change: Advances in communication, transportation, science, and medicine have made significant changes in people's lives, especially in developed nations; however, these changes also have created the potential for new disasters, ranging from global warfare to localized technological disasters at toxic waste sites.

D. Social Institutions and Change: During the past twentieth century, many changes occurred in the family, religion, education, the economy, and the political system.

E. Changes in physical environment, population, technology, and social institutions operate together in a complex relationship, sometimes producing consequences we must examine by using our sociological imagination.

F. A few final thoughts: One purpose of this text is to facilitate understanding of different viewpoints in order to resolve social issues, thereby, producing a better way of life not only in this country but worldwide.

ANALYZING AND UNDERSTANDING THE BOXES

After reading the chapter and studying the outline, re-read the boxes and write down key points and possible questions for class discussion.

Sociology and Everyday: Life How Much Do You Know About Collective Behavior and Environmental Issues?

Key Points:

Discussion Questions:

1.

2.

3.

Sociology in Global Perspective: "Flash Mobs": Collective Behavior in the Information Age

Key Points:

Discussion Questions:

1.

2.

3.

You Can Make a Difference Its Now or Never: the Imperative of Taking Action Against Global Warming.

Key Points:

Discussion Questions:

1.

2.

3.

PRACTICE TESTS

MULTIPLE CHOICE QUESTIONS

Select the response that best answers the question or completes the statement:

1. __________ is the alteration, modification, or transformation of public policy, culture, or social institutions over time. (p. 482)
   a. Social movements
   b. Social stratification
   c. Collective behavior
   d. Social change

2. All of the following are factors that contribute to the likelihood that collective behavior will occur, *except*: (pp. 480-481)
   a. the presence of deviant behavior.
   b. structural factors that increase the chances of people responding in a particular way.
   c. timing.
   d. a breakdown in social control mechanisms and a corresponding feeling of normlessness.

3. A ______ is a number of people who share an interest in a specific issue but are not in close proximity to each other. (p. 483)
   a. crowd
   b. mass
   c. riot
   d. mob

4. Crowds, mobs, riots, panics, fads, fashions, and public opinion are examples of __________. (p. 480)
   a. organizational behavior
   b. institutional behavior
   c. social movements
   d. collective behavior

5. Casual crowds are: (p. 484)
   a. comprised of people who specifically come together for a scheduled event and thus share a common focus.
   b. situations that provide an opportunity for the expression of some strong emotion.
   c. comprised of people who happen to be in the same place at the same time.
   d. comprised of people who are so intensely focused on a specific purpose or object that they may erupt into violent or destructive behavior.

6. Rachel Carson's classical research *Silent Spring* made people aware of: (p. 481)
   a. love canal
   b. hazards of chemicals in food
   c. mob activity
   d. none of the above

7. A common stimulus is an important factor in collective behavior. For example, Rachel Carson's book Silent Spring, which detailed the dangers of __________, led many people and groups to turn to activism.(p. 480)
   a. overpopulation
   b. pesticides
   c. nuclear war
   d. air pollution

8. When the residents of Love Canal burned both the governor and the health commissioner in effigy, they were engaging in a/an: (p. 485)
   a. acting crowd.
   b. casual crowd.
   c. conventional crowd.
   d. panic.

9. A(n) __________ is a relatively large number of people who mutually transcend, bypass, or subvert established institutional patterns and structures. (p. 480)
   a. group
   b. cohort
   c. aggregate
   d. collectivity

10. Which of the following contributes to the likelihood that collective behavior will occur? (p. 480)
    a. structural factors that increase the chances of people responding in a particular way
    b. timing
    c. a breakdown in social control mechanisms
    d. (all of the above)

11. People are more likely to engage in collective behavior in response to a common stimulus if _________. (p. 481)
    a. they are gathered together in one location
    b. they are physically separated from one another
    c. they act through official channels
    d. social control mechanisms are strong and effective

12. A(n) __________ is a relatively large number of people who are in one another's immediate vicinity. (p. 483)
    a. crowd
    b. mass
    c. institution
    d. society

13. A(n) __________ is a number of people who share an interest in a specific idea or issue but who are not in one another's immediate vicinity. (p. 483)
    a. crowd
    b. mass
    c. institution
    d. society

14. Convergence theory focuses on: (p. 486)
    a. the social psychological aspects of collective behavior, including how moods, attitudes, and behavior are communicated.
    b. how social unrest is transmitted by a process of circular reaction.
    c. the importance of social norms in shaping crowd behavior.
    d. the shared emotions, goals, and beliefs many people bring to crowd behavior.

15. All of the following statements regarding the emergent norm theory are true, *except*: (p. 487)
    a. Emergent norm theory is based on the symbolic interactionist perspective.
    b. Sociologists using the emergent norm approach seek to determine how individuals in a given collectivity develop an understanding of what is going on, how they construe these activities, and what types of norms are involved.
    c. Emergent norm theory points out that crowds sometimes are irrational.
    d. Emergent norm theory originated with sociologists Ralph Turner and Lewis Killian.

16. Rumors, gossip, fashions, and fads are examples of _____ behavior. (p. 488)
    a. mob
    b. mass
    c. irrational
    d. casual

17. A form of dispersed collective behavior that occurs when a large number of people react with strong emotions and self-destructive behavior to a real or perceived threat is known as: (p. 489)
    a. mob behavior.
    b. mass behavior.
    c. mass hysteria.
    d. contagious behavior.

18. Orson Wells's famous radio dramatization, *The War of the Worlds*, is an example of: (p. 489)
    a. flash mob.
    b. riot.
    c. fad.
    d. panic.

19. ________ consists of the attitudes and beliefs communicated by ordinary citizens to decision makers. (p. 490)
    a. public opinion
    b. fashion statements
    c. mob behavior
    d. public mass

20. According to the text, ________ is information provided by individuals or groups that have a vested interest in furthering their own cause or damaging an opposing one. (p. 491)
    a. propaganda
    b. public opinion
    c. political rhetoric
    d. a press release

21. Social movements are more likely to flourish in which of the following types of society? (p. 492)
    a. totalitarian
    b. authoritarian
    c. socialist
    d. democratic

22. According to emergent norm theory, individuals act _______ when they are part of a crowd. (p.487)
    a. rationally
    b. irrationally
    c. without any regard to social norms
    d. violently

23. Grassroots environmental movements are an example of: (p. 493)
    a. reform movements.
    b. alternative movements.
    c. dissident movements.
    d. revolutionary movements.

24. According to relative deprivation theory: (pp. 495-496)
    a. people who are satisfied with their present condition are more likely to seek social change.
    b. certain conditions are necessary for the development of a social movement.
    c. people who feel that they have been deprived of their "fair share" are more likely to feel that change is necessary and to join a social movement.
    d. some people bring more resources to a social movement than others.

25. ______ theory is based on the assumption that six conditions, including structural conduciveness and structural strain, must be present for the development of a social movement. (p. 496)
    a. Value-added
    b. Relative deprivation
    c. Resource mobilization
    d. Emergent norm

26. ______ theory focuses on the ability of members of a social movement to acquire resources and mobilize people in order to advance their cause. (p. 496)
    a. Value-added
    b. Relative deprivation
    c. Resource mobilization
    d. Emergent norm

27. ______ is a social movement based on the belief that patriarchy is a root cause of environmental problems. (p. 498)
    a. Ecology Today
    b. Conflict Ecologists
    c. Environmental Justice
    d. Ecofeminism

28. According to sociologist Steven Buechler, social movements are historical products of the _________. (p. 499)
    a. age of reason
    b. age of modernity
    c. age of industrialization
    d. agrarian age

29. The belief that a disproportionate number of hazardous facilities are placed in low-income areas populated by people of color is known as: (p. 498)
    a. environmental racism.
    b. environmental justice.
    c. reverse environmentalism.
    d. racial pollution.

30. All of the following statements regarding natural disasters are true, *except*: (pp. 500-502)
    a. Major natural disasters can dramatically change the lives of people.
    b. Trauma that people experience from disasters may outweigh the actual loss of physical property.
    c. Natural disasters are not affected by human decisions.
    d. Disasters may become divisive elements that tear communities apart.

TRUE-FALSE QUESTIONS

T F 1. Collective behavior lacks an official division of labor, hierarchy of authority, and established rules and procedures. (p. 480)

T F 2. People are more likely to act as a collectivity when they believe it is the only way to fight those with greater power and resources. (p. 483)

T F 3. People gathered for religious services and graduation ceremonies are examples of casual crowds. (p. 484)

T F 4. Panics often arise when people believe that they are in control of a situation. (p. 485)

T F 5. Protest crowds engage in activities intended to achieve specific political goals. (p. 485)

T F 6. Sociologist Robert E. Park was the first U.S. sociologist to investigate crowd behavior. (p. 486)

T F 7. According to Clayman, people like to "boo" a speech independently. (p. 487)

T F 8. For mass behavior to occur, people must be in close proximity geographically. (p. 487)

T F 9. Public opinion does not always translate into action by decision makers in government and industry or by individuals. (pp. 490-491)

T F 10. Mothers Against Drunk Driving is an example of a reform movement. (p. 493)

T F 11. A resistance movement is one that embraces social change. (p. 494)

T F 12. Revolutionary movements also are referred to as expressive movements. (p. 494)

T F 13. Movements based on relative deprivation are most likely to occur when people have unfulfilled rising expectations. (p. 495)

T F 14. Sociologist Kai Erickson has suggested that small, natural disasters do not affect people in a meaningful way. (p. 500)

T F 15. One purpose of this text is to expose the readers to different viewpoints. (pp. 504-505)

SOCIOLOGY IN OUR TIMES: DIVERSITY ISSUES

1. Have you ever participated in a social movement? If so, what types of issues were involved in the movement? Were the issues related to race/ethnicity, class, gender, age, sexual orientation, religion, or disability? Why are so many social movements intertwined with one or more of these characteristics?

2. Watch television news programs for examples of collective behavior or social movements. Who are the participants? What are the major issues involved? When did the behavior occur? Where did it occur? Why did participants engage in this behavior? Does the behavior of participants have any impact on your life? Why or why not?

3. Can you cite evidence that environmental racism exists in your community or city? Why do some sociologists believe that environmental justice is a pressing social issue?

4. Based on your race/ethnicity, gender, class, and age, how do you think you will be affected by changes in the physical environment, population, technology, and social institutions in the future?

INTERNET EXERCISE

Access the Internet and use the search engine of your choice to find examples of different behaviors of groups and the activities in which they participate:

1. Civil disobedience
2. Crowd
3. Mass behavior
4. Mob

INFOTRAC COLLEGE EDITION ONLINE READINGS AND EXERCISES

You will need to access the Infotrac College Edition to answer the following questions. Once logged on, type ""Crowd lines up outside Akron, Ohio " in the **keyword** search field. Next click on "Crowd lines up outside Akron, Ohio-area Chick-fil-a for a year of free food." Read the selection and answer the following:

a. Why did the crowd begin accumulating before sunup?

b. What were some of the reasons given (besides free food) for the crowd to ever accumulate? How do the people feel when they are apart of something like this experience?

Return to the **keyword** search field and type in "Breaking the Iron Law of Oligarchy: Union Revitalization in the American Movement." Read the article and answer the following questions:

a. How was the labor union social movement able to break out of bureaucratic conservatism?

b. What is the impact of leader's concerns with organizational maintenance?

c. Were the bureaucratic changes democratic in nature?

STUDENT CLASS PROJECTS AND ACTIVITIES

1. The uprising against socialism in the Soviet Union in 1991 is an example of a revolutionary movement. It sought, and apparently produced an entire restructuring of the Soviet political and economic system by overthrowing the existing social order and creating a new one. This could be considered the second of the revolutions that have occurred in this country since the Russian Revolution of 1917. Choose one of these two revolutions and trace the development of the movement. Include the conditions necessary for the development of the movement, including the conditions necessary for the development of a social movement according to Smelser, as mentioned in the text. In your paper, provide examples of the six conditions of the revolution and critique the success of the movement. Submit your paper to your instructor at the appropriate time, following any additional instructions you have been given.

2. Collect articles dealing with some form of collective behavior as discussed in the text. The articles could focus on fads, fashions, mobs, the public, or specific types of social movements. Provide a copy of each article in your paper. You are to (1) summarize each article; (2) explain the significance (if any) of the article; (3) write a personal reaction to each article; (4) provide a conclusion and personal evaluation of the project; and (5) provide a bibliographic reference for each article selected. You should submit your paper to your instructor at the appropriate time, following any additional instructions given to you.

3. Write a paper on the impact of artificial intelligence in our society today and upon the societies for tomorrow. Provide at least three bibliographical references in your paper. Submit your paper to your instructor at the appropriate time, following any additional instructions you may have been given.

ANSWERS TO THE PRACTICE TESTS, CHAPTER 16

Answers to Multiple Choice Questions

1. d This is something that is continually occurring at a variety of levels. (p. 480)
2. a All of the following are factors that contribute to the likelihood that collective behavior will occur, *except*: the presence of deviant behavior. (pp. 480-481)

3. b A mass is a number of people who share an interest in a specific idea or issue but who are not in one another's immediate vicinity. (p. 483)
4. d A variety of venues and values attached to collective behavior. (p. 480)
5. c Casual crowds are comprised of people who happen to be in the same place at the same time. (p. 484)
6. b Carson's book, *Silent Spring,* documents the hazards of chemicals in our food supply and the destruction of wildlife. (p. 481)
7. b We see this in cancer clusters throughout the Midwest where spraying occurs on crops. (p. 480)
8. a When the residents of Love Canal burned both the governor and the health commission in effigy, they were engaging in an acting crowd. (p. 485)
9. d Notice of differences in definitions, especially this one which would probably be referred to as a social group by most Americans. (p. 480)
10. d All of the above apply. (p. 480)
11. a For logistics reasons, it makes it easier. (p. 481)
12. a A standard definition; begin to compare these definitions with those other types of groups. (p. 483)
13. b A standard definition, begin to compare these definitions with those other types of groups. (p. 483)
14. d Convergence theory focuses on the shared emotions, goals, and beliefs many people bring to crowd behavior. (p. 486)
15. c All of the following statements regarding the emergent norm theory are true, *except*: emergent norm theory points out that crowds sometimes are irrational. (p. 487)
16. b Rumors, gossip, fashions, and fads are examples of mass behavior. (p. 488)
17. c Mass hysteria is a form of dispersed collective behavior that occurs when a large number of people react with strong emotions and self-destructive behavior to a real or perceived threat. (p. 489)
18. d Welles' famous radio dramatization is an example of a panic. (p. 489)
19 a the attitudes and beliefs communicated by ordinary citizens to decision makers is known as public opinion (p. 490)
20. a According to the text, propaganda is information provided by individuals or groups that have a vested interest in furthering their own cause or damaging an opposing one. (p. 491)
21. d Where freedoms allow expression. (p. 492)
22. a This idea has gone back and forth throughout sociological history. Look at the history of rationalism, and there is not consensus on this. (p. 487)
23. a Grassroots environmental movements are an example of reform movements. (p. 493)
24. c According to relative deprivation theory, people who feel that they have been deprived of their "fair share" are more likely to feel that change is necessary and to join a social movement. (pp. 495-496)
25. a Value-added theory is based on the assumption that six conditions, including structural conduciveness and structural strain, must be present for the development of a social movement. (p. 496)

26. c Resource mobilization theory focuses on the ability of members of a social movement to acquire resources and mobilize people in order to advance their cause. (p. 496)
27. d Ecofeminism is based on the belief that patriarchy is a root cause of environmental problems. (p. 498)
28. b According to Steven Buechler, social movements are historical products of the age of modernity. (p. 499)
29. a The belief that a disproportionate number of hazardous facilities is placed in low-income areas populated by people of color is known as environmental racism. (p. 498)
30. c All of the following statements regarding natural disasters are true, *except*: natural disasters are not affected by human decisions. (pp. 500-502)

Answer to True-False Questions

1. True (p. 480)

2. True (p. 483)

3. False People gathered for religious services and graduation ceremonies are examples of conventional crowds. (p. 484)

4. False Panics may arise in response to events that people believe are beyond their control. (p. 485)

5. True (p. 485)

6. True (p. 486)

7. False According to Clayman, people wait to coordinate their booing with other people; they do not wish to "boo" alone. (p. 487)

8. False Mass behavior often takes place when people who are geographically separated from one another respond to the same event in much the same way (for example, rumor, fashion, or fad). (p. 487)

9. True (pp. 490-491)

10. True (p. 493)

11. False A resistance movement is one that does not embrace social change, but rather seeks to prevent change. (p. 494)

12. False Religious movements also are referred to as expressive movements. (p. 494)

13. True (p. 495)

14. False Kai Erickson has suggested that even "small" natural disasters change the lives of many people. (p. 500)

15. True (pp. 504-505)